Decision Support Systems for Wastewater Facilities Management

by:

Z. Cello Vitasovic, P.E.
DHI Inc.

2004

WERF

The Water Environment Research Foundation, a not-for-profit organization, funds and manages water quality research for its subscribers through a diverse public-private partnership between municipal utilities, corporations, academia, industry, and the federal government. WERF subscribers include municipal and regional water and wastewater utilities, industrial corporations, environmental engineering firms, and others that share a commitment to cost-effective water quality solutions. WERF is dedicated to advancing science and technology addressing water quality issues as they impact water resources, the atmosphere, the lands, and quality of life.

For more information, contact:

Water Environment Research Foundation
635 Slaters Lane, Suite 300
Alexandria, VA 22314-1177
Tel: (703) 684-2470
Fax: (703) 299-0742
www.werf.org
werf@werf.org

This report was co-published by the following organizations. For nonsubscriber sales information, contact:

Water Environment Federation
601 Wythe Street
Alexandria, VA 22314-1994
Tel: (800) 666-0206
Tel: (703) 684-2452
Fax: (703) 684-2492
www.wef.org
pubs@wef.org

IWA Publishing
Alliance House, 12 Caxton Street
London SW1H 0QS, United Kingdom
Tel: +44 (0) 20 7654 5500
Fax: +44 (0) 20 7654 5555
www.iwapublishing.com
publications@iwap.co.uk

Library of Congress Catalog Card Number: 2004103304

WEF ISBN: 1-57278-193-9 IWAP ISBN: 1-84339-720-X

DHI Inc.

The research on which this report is based was funded, in part, by the United States Environmental Protection Agency through Cooperative Agreement No. CR-827345-01 with the Water Environment Research Foundation (WERF). Unless an EPA logo appears on the cover, this report is a publication of WERF, not EPA. Funds awarded under the Cooperative Agreement cited above were not used for editorial services, reproduction, printing, or distribution.

ACKNOWLEDGMENTS

Report Preparation

Principal Investigator:
Dr. Z. Cello Vitasovic, P.E., *DHI Inc.*

Project Team:

Dr. Michael Barnett, *Gensym Corporation*

Elizabeth Davidson, *Davidson & Associates*

David Duest, *MWRA*

Dick Finger, *King County, Washington*

Leonard Gipson, *Philadelphia Water Department*

Michael Sweeney, *EMA Inc.*

Project Subcommittee

Tyler Richards, Chair, *Gwinnett County, Georgia*

C. Michael Bullard, P.E., *Hazen & Sawyer*

Dr. Robert D. Hill, P.E., *EMA Inc.*

Beverly J. Ingram, *American Water Services*

Michael Marcotte, P.E., DEE, *Department of Public Works, City of Houston, Texas*

Wiff Peterson, *Aquarion Services*

Richard Ottman, P.E., *Metropolitan Council Environmental Services*

Water Environment Research Foundation Staff

Director of Research: Daniel M. Woltering, Ph.D.
Project Manager: Roy Ramani, Ph.D.

ABSTRACT AND BENEFITS

Abstract:

Municipalities not only have a need for purely transactional systems (accounting, personnel, billing, etc.), but they also need to integrate the sophisticated analytical tools that are used to address the scientific or engineering issues that are very important for the performance of wastewater utility. Most commercially available systems are structured to function within the transactional paradigm and may be inadequate for full range of business processes encountered in a wastewater utility.

Wastewater utilities are overwhelmed with data. Yet often, managers do not have enough information to efficiently support many of their business functions. Although significant investments into information technology have been made in the last two decades, decision makers often cannot access their organizationís information. Decision analysis provides the conceptual framework and specific tools for describing the decision-making process. The framework provides methods and a standard notation for presenting decision-making processes in a structured way. This project describes decision support concepts, examines how private industry has approached this problem, and discusses how this technology could be applied to improve the management of wastewater facilities.

Benefits:

- Assists decision maker to better use their information technology (IT) resources to address their business needs.
- Highlights the diversity and complexity of decision support system applications.
- Describes how investments into IT are best leveraged to improve the management of wastewater agencies.
- Defines the specific functional requirements for DSS that would work well for a wastewater agency.
- Describes decision support systems and identifies their components.
- Demonstrates how private industry addresses issues related to decision support.
- Reviews practices related to decision support in wastewater industry.

Keywords: system integration, business processes

TABLE OF CONTENTS

LIST OF FIGURES

LIST OF ACRONYMS

CAD	Computer-Aided Drafting
CASE	Computer Aided Software Engineering
CIP	Capital Improvement Program
CMMS	Computerized Maintenance Management Software
DA	Decision Analysis
DATA	Decision Analysis software from Tree Age
DPL	Decision Programming Language
DSS	Decision Support System
GIS	Geographic Information System
GUI	Graphical User Interface
ID	Identification
IT	Information Technology
LIMS	Laboratory Information Management System
SCADA	Supervisory Control and Data Acquisition
SCOR	Supply Chain Operations Reference
TOM	Telecommunications Object Map
URL	Uniform Resource Locator
WERF	Water Environment Research Foundation

Executive Summary

Effective management of a wastewater utility includes the application of a range of business processes, from accounting to science and engineering. Business processes are executed on operational, tactical, and strategic levels, depending on the time frame and scope of the decision. Strategic decisions have a long-term time frame and a broad scope, whereas operational decisions at the other extreme have short time frame (sometimes online) and a specific and narrow scope.

Decision support systems (DSS), as defined in this research, are software applications that collect and organize data from several different applications throughout the enterprise. DSS address issues that extend beyond a single group or functional domain. DSS aim to convert data into knowledge (usable and actionable information).

ES.1 Decision Making and Analysis

Decision making includes three main components: data, tools, and business processes. DSS can bring improvements in each of these components by:

- Providing methods for analyzing the business processes and representing the business processes in a formalized way
- Developing and integrating tools that process the data and help the users (from operators to managers) transform raw data into usable, actionable information
- Integrating data that exists in many separate systems that do not easily communicate with each other

There are two main types of decision analysis: descriptive and prescriptive. Descriptive decision analysis includes investigation of how human beings or groups actually make decisions. Prescriptive decisions analysis examines the processes that can be used to improve the quality of the decisions that people make. Although both of these approaches can provide valuable insight, prescriptive decision analysis is used more directly in the development and implementation of DSS.

Typically, applied decision analysis has as its basis the principle of consequence: decision makers define the possible approaches and outcomes, and evaluate different alternatives against the defined objectives. Decision analysis theory provides methods and tools to map out and evaluate the different paths and different outcomes.

ES.2 DSS in the Wastewater Industry

Private industry has made significant progress in applying DSS to business improvement at various organizational levels. The development of generic business reference models such as

SCOR (Supply Chain Operational Reference; 2000) or TOM (Telecommunications Operations Map; 2000) are especially interesting examples that have no parallel or equivalent in the wastewater industry.

Decision analysis, decision support systems, and system integration have not yet been widely used or accepted in the wastewater industry. Examples of some early applications of these concepts to wastewater industry are presented in appendices to this report, which include contributions (position papers and case studies) from companies that are engaged in this work in the United States. The lag in application of these concepts between private (parallel) industries and the wastewater industry is primarily caused by strong profit pressures and incentives that are driving the private industry, but that are generally lacking in wastewater industry.

System integration is closely linked with the concept of DSS. Many wastewater agencies have invested significant resources into automation, and have developed and/or implemented a number of computer systems to handle specific aspects of their business. In most cases, each individual system addresses a specific functional domain and few links exist between different computer systems. Typically, there are significant barriers to exchanging information across functional and organizational boundaries.

On a practical and immediate level, this research presents the following suggestions and thoughts to a manager in a wastewater agency:

- Wastewater municipalities could benefit from implementing some tools and/or methods from the field of decision analysis. Business process analysis is presented as a specific example of an activity that can lead to improvements in operation.
- Integration of (existing and future) computer systems can introduce a whole new level of functionality and enhance the decision-making processes on several levels. Integration can also help the organization maximize the return on the investments that are made into information technology.
- The process of applying technology to solve a business problem needs to be based on a thorough analysis of the business problem. Understanding the business needs is critical to making the right choices from the bewildering multitude of technological choices.

Appendices A through C for this research are available on the WERF Website and/or the third-party publishers listed on page ii and the back cover of this report. For more information, call the organization from which you purchased this report or contact WERF at 703-684-2470 or visit the WERF Website at www.werf.org.

CHAPTER 1.0

INTRODUCTION

This research project was initiated because many Water Environment Research Foundation (WERF) subscribers (especially municipalities) felt that something was missing when it came to the application of information technology (IT) to addressing their business needs. The three primary issues that the managers in municipalities were experiencing may be summarized as follows:

Computerization and automation. In the last 20 years or so, municipalities have invested significant amounts of money into computerization and automation. Many systems have been implemented to support different business processes and units across the organization. There is a sense in the user community, however, that the expected benefits from these investments may not have been realized. The specific concern is that applications do not easily share data, and that they appear as isolated islands of information on a map of organizationís information technology infrastructure.

Volume of data. Introduction and implementation of different computer systems have resulted in agencies collecting an increasing, often overwhelming, volume of data. However, there is also a belief that this data is not accessible, and that it is ì locked upî in different computer systems that exist throughout the agency. There also arises a sense of ì data overload and information deficiencyî because, in spite of the staggering volume of the collected data, meaningful information may be difficult to obtain. Tools and procedures for quality control of data are often lacking when the volume of data is large; and this may erode confidence in using data for business decision making.

Decision support systems. In the last few years, the term decision support systems (DSS) has been used with increased frequency, but often in reference to different products or services. This has caused confusion: there was thus also a desire to define what this term means, and provide some guidance it terms of distinguishing between different products and/or services.

Too frequently, especially when computer system professionals are involved, a discussion about decision support is soon narrowed down to the selection or application of a yet newer generation of computer systems. This approach might neglect the real business issues or the complex nature of the decision-making process. On the other hand, a purely theoretical discourse on the decision-making process is not very helpful to a practicing manager or engineer at a municipality. To address broader issues of decision making and more fully respond to the three main concerns listed above, this research will attempt to keep a proper balance between a conceptual framework and technology issues related to delivery of practical solutions.

1.1 Research Framework

The focus of the research, and this report, starts on a more theoretical, broader, and conceptual level and then gradually moves to more specific and applied topics.

In addition to addressing theoretical decision support issues, this project will also try to answer the following practical question: how can investments into information technology best be leveraged to improve the management of wastewater agencies?

To provide a more complete answer to this question, it is useful to take a step back and examine what exactly the information systems are supposed to do, and how they can add value to different business processes. A decision is normally the result of a *business process*; a conceptual view of the broader issue is presented below:

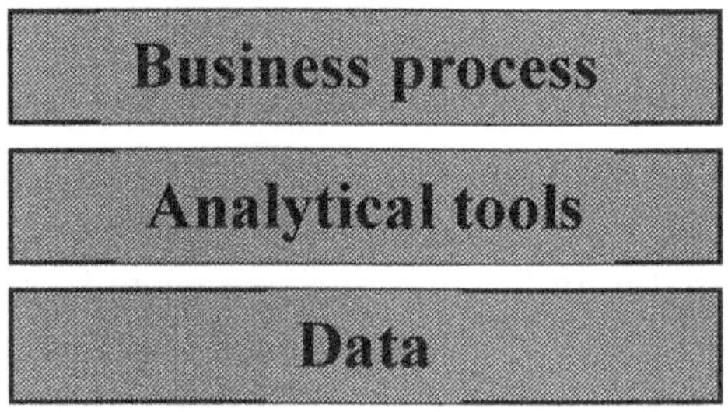

Figure 1-1. ì Three-Layerî View of Business Processes.

This ì three-layerî view is used to examine different issues throughout this report; and most of the information is organized around these three components. The ì Dataî box in the diagram above indicates all the information that is necessary to make a business decision, and the analytical tools are modules that analyze/process this data and make it available to the decision maker(s).

The top box represents a business process. This critical element includes the activities and decisions that can occur on three levels:

1. Operational. The decisions on this level are short term, often online and automated (e.g., adjusting process control set points and manipulating final control elements (e.g., gates, pumps).
2. Tactical. Decisions and issues on this level have a bit longer time constant, measured in terms of days or weeks. Examples include scheduling crews, planned (daily/weekly) maintenance activities, sensor calibration, preparation and submittal of reports, response to emergency conditions, etc.
3. Strategic. On this level, issues and decisions have the longest time constant, measured in months or years. Strategic level includes longer term planning for resources (e.g., staff, equipment, spare parts, or equipment purchase) and facilities (pumping stations, control structures, etc.).

A DSS can facilitate the decision making on one or more of the levels outlined above. A successful strategy for the development of DSS might include breaking up the problem into three parts to address the different levels of decisions.

Finally, the appendices provide the reader with an overview of the current efforts related to decision support and system integration both in private industries and in the wastewater industry. Several leading U.S. companies that provide services related to decision support and/or system integration were asked to contribute papers and materials that present their vision and also case studies. The research team is grateful for these contributions.

1.2 Report Guide

This section provides an overview of issues that are addressed in this report. This research project is challenging for the following reasons:

- The scope of this research is broad. Wastewater utility management includes a wide of decision-making processes on several levels (strategic, tactical, operational). Decisions may range from selecting a site for a treatment plant, setting the recycle flow rate in an activated sludge treatment plant, establishing maintenance priorities and schedules, or managing project budgets. To address these diverse issues in a consistent and coherent way, it is necessary to adopt a generic approach that makes sense across the broad range of business needs and issues.
- There are significant differences in the need for integration between different municipalities. Some agencies have already implemented many (specific) automated systems and computer applications, and thus they see an immediate need for integration. Some other municipalities are only now considering automation and computer tools: and although some of them view integration as ì something we may have a need for in the future,î there are benefits from considering integration issues early in the process of automation. The content of this report will hopefully provide useful information to municipalities on both sides of the spectrum.
- In many municipalities, specific business processes have been addressed by the stand-alone and separate IT solutions that address a specific business domain such as maintenance management, laboratory, process control, or mathematical models. Generally, these systems were developed to address specific business needs of groups that have a narrow business focus. There are many examples of such systems, and these ì local needsî have been well researched and addressed. System-wide and/or integrated solutions, however, are only beginning to emerge in municipalities and thus the report cannot easily point to a few established tools that one could compare and make recommendations.
- The term ì decision support systemî has been used to describe systems that vary greatly in scope, functionality, and architecture. Taken to the extreme, any computer system that helps perform a business process or make (even simple decisions), might be referred to as a DSS. For the purpose of this research, a DSS is assumed to span across more than just one functional domain, and to support decisions that are reasonably complex in terms of data requirements and the level of analysis involved.

The intended audience for this report is professionals whose expertise is in wastewater utility management, and not information technology (IT). As much as possible, the focus of this report is on improving the tools for wastewater utility management, and technology is discussed mainly as a way to accomplish this.

Different readers, depending on their expertise domain and their role in the organization, may be interested in different aspects of this research. Although hopefully each person might benefit from reading the entire report, here are some thoughts on how the content of this report may address different issues:

The main body of the report starts with a definition of the business issues that led to this research; this information should be of interest to most readers. The chapter on Decision Analysis Science may be helpful to those who would like a quick overview of decision analysis. A number of references are included for those who desire more detail. This chapter also provides an introduction to the main ideas of decision analysis, and although it might seem ìtoo theoreticalî at a first glance, it should help one understand the overall framework for decision making. The chapter on Defining the Problem and the Objectives should be useful to all readers, because this activity is an essential part in defining DSS scope and functionality.

The chapter titled, ìHow is Decision Analysis Related to DSS,î is intended to help bridge the gap between the generic and theoretical topic of Decision Science and the pragmatic and specific goals of building systems that will help manage a wastewater utility. This section is intended to show how the theory of decision analysis fits into the specific problem. For a manager who is not interested in the details of decision analysis science, this section briefly describes and summarizes practical applications of decision analysis to managing a wastewater agency.

The chapter on Context for DSS discusses the process/methodology for approaching DSS development. The chapter on IT Aspects of DSS discusses issues related to DSS technology and how a DSS fits into an overall IT infrastructure in a typical wastewater agency. This chapter discusses the ìhowî of building a DSS. The chapter on Functional User Requirements is intended to describe what a DSS might look like, and what it may be able to do for different types of users. The chapter on Technical Requirements provides a summary of technical requirements for a DSS. Finally, the final chapter summarizes the content of the report.

CHAPTER 2.0

THE DECISION-MAKING PROCESS: INTRODUCTION TO DECISION ANALYSIS

Individuals, groups, and organizations are continuously engaged in decision-making processes. Individuals make decisions (on many levels) so frequently that sometimes the entire process may seem instinctual, like breathing or walking. When decisions are made by groups or organizations, the process may often seem very complex, as it can be influenced by many conflicting forces, agendas, goals, and uncertainties. The effort to describe in a rigorous and structured way ì how the decisions really get madeî may seem truly daunting. Decision science and decision analysis examine the nature of such decision-making processes, ultimately seeking to identify, understand, and improve the underlying rules and models.

Decision analysis includes two areas of research and practice:

- Descriptive decision analysis investigates how people (individuals and/or groups) think and make decisions. The goal is to truly understand the underlying thought processes upon which decisions are made.
- Prescriptive decision analysis investigates how people *should be* making decisions: it examines the procedures, rules, and methods for selecting the best alternative(s) and achieving the best possible result. The goal of the prescriptive decision analysis is to provide methods, tools, and systems that people can use to formulate the problem, identify available alternatives, structure the problems, evaluate possible outcomes, and ultimately choose the most promising alternative.

2.1 Descriptive Decision Analysis

The descriptive decision analysis is a broad research area that relies heavily on work in psychology and psychiatry. Although descriptive decision analysis is not the focus of this project, some aspects of this field might provide a meaningful insight to managers and executives to help understand the mechanisms that people use to arrive at decisions.

The detailed psychology and psychiatry aspects of descriptive decision analysis are quite complex and a full investigation was not within the scope of research on this project. Some interesting lessons can be learned, however, about how people perceive the world around them and how they make decisions. Excellent introduction to this field is provided by March (1994) and Baron (2000). Additionally, a number of texts on psychology and the ì thinking processî written for lay persons (e.g., Sachs, 1987) are very readable and accessible, and may provide a keen insight to such processes.

The research team would like to point out two themes within descriptive decision analysis that warrant consideration and thought.

2.1.1 Rational vs. Rule-based Decision Making

Prescriptive decision analysis is based on the *logic of consequence* (March, 1994): this model of decision making assumes that alternative paths (solutions) are evaluated and compared against each other based on their impact on the defined objectives. It presents the problem of decision making as a search for the best possible solution.

In everyday life, the logic of consequences is not necessarily how people make decisions. An entirely different model follows the *logic of appropriateness*: within this model, people make decisions based on rules. The rules for decision making within this model are based on three components:

1. Recognition: the person (decision maker) recognizes the type of a situation.
2. Identity: the decision maker considers his own identity. The identity includes a set of rules for making decisions, and it includes specific identities (e.g., manager, operator, engineer, doctor, father, soldier).
3. Rules: each of the roles/identities carries with it a set of predetermined rules of behavior. The decision maker would consider the question: what is a person like me supposed to do in a situation like this?

The rule-based approach is very common, and many aspects of our lives are based on such rules. Our educational system instills these rules, and teaches people how to recognize the situations.

The methods from the prescriptive decision analysis are based on the logic of consequence. This fits well into a structured framework that can be implemented to evaluate different alternatives, and can improve our decision-making process. However, the logic of consequence has its limitations.

2.1.2 The Importance of Synthesis

In his 1987 book, *The Man Who Mistook His Wife for a Hat*, neurologist Oliver Sachs describes a number of cases involving patients whose ability to perceive has been affected by neurological damage. It turns out that the brain processes the visual information in two ways: examining the details and then synthesizing the details into a coherent visual image that makes sense. These two related but distinct functions are processed in different parts of the brain. The patient described in the book title still had the full ability to see the details, but had lost the ability to make sense out of it all.

The ì synthesisî ability allows us to recognize the visual images even if provided with a rough outline or few details. The brain ì fills in the blanksî and, therefore, we can recognize a person from merely a silhouette, or from a caricature that uses only a few key details. If all we can register is the details, we can no longer recognize a person after they got a haircut or got rid of their eyeglasses.

In some ways, many organizations have similar symptoms. Some reasonable tools are in place to deal with the details, yet this is not sufficient. The role of the DSS is to synthesize all the data into information and make sense out of it. In most IT systems today, synthesis is currently lacking. This is where DSS should helpó DSS should aim to convert data into knowledge (usable and actionable information).

2.2 Prescriptive Decision Analysis

Prescriptive decision analysis is more directly applicable to this research, but still all aspects of this fairly broad and developed field cannot be considered in full detail within this research and this report. The main goal of this research is to provide the context that would allow professionals in the wastewater industry to improve their decision-making processes, and the level of detail included in the report (regarding the theoretical underpinnings) is therefore gauged to satisfy these pragmatic goals. A brief introduction of the methods used for prescriptive decision analysis is included below.

Some decisions are obviously much harder than others. Decisions can be hard for the following reasons:

- Alternatives: The effect on the outcome of the choice between alternatives is not easily understood. The two main reasons for this are:
  - Complexity of the problem: the problem may include a number of different components/aspects/processes that interact and influence each other.
  - Uncertainty: there may not be access to information that would have a profound impact on the best choice for a preferred alternative. Often the choice must be made without such information.
- Multiple, possibly competing, objectives: Often, different stakeholders may have different objectives and perspectives, and they may also differ in their assessments and assignment of risk. Therefore, a framework must be put in place that allows decision making to occur with a balanced and structured approach that considers such situations.

Making a decision includes recognizing the problem, considering and evaluating the available alternatives, and selecting one alternative. A good decision is one that is best suited for achieving a desired outcome.

Achieving the desired outcome does not necessarily mean that the decision was good. An example often used to demonstrate this concept is when somebody wins the lottery: although the desired outcome was achieved, the decision to invest money into purchasing the lottery ticket may not have been the best decision.

The process for making a decision could be broken down into the following components:

- defining the problem and the objectives;
- identifying alternatives to address the problem;
- developing models to assess the impact of each alternative on the achievement of the objectives; and

- implementing the preferred alternative.

Decision analysis provides us with tools and methods for addressing each aspect of decision making.

2.2.1 Defining the Problem and the Objectives

A popular saying goes: ìIf you donít know where you are going, any road will get you there.î Properly defining the objectives is a critical step in the decision-making process, and although there are differences in approaches to decision analysis, there seems to be universal agreement that the definition of objectives is the critical first step. We must begin with the end in mind.

The objectives are normally identified through interviews and workshops with the different people that are in some important way related to the decision-making process. Once the objectives are identified, they need to be placed into a structure, based on how they relate to each other. Decision analysis distinguishes between two types of objectives: fundamental objectives and means objectives.

A good question to ask is: ìwhy is this objective important?î If the answer is clear and pragmatic, (e.g., ìto reduce the cost of the activityî), this is most likely a means objective. If the best answer is ìjust because it is important,î this is most likely a fundamental objective.

Fundamental objectives are normally organized in hierarchies. Means objectives are often structured and represented in networks.

The process of identifying the objectives can be helped by a skilled facilitator who can ìextractî the required information through the interviews, discussions, and workshops. Keeney (1992) offers several ìdevicesî to such facilitators, including:

- Preparing a ìwish listî from each key stakeholder
- Outlining alternatives that will stimulate thinking about objectives
- Asking about current issues and problems the decision maker is facing
- Identifying consequences and asking if they are acceptable
- Using goals, constraints, and existing guidelines (goal is defined as a specific measure of an objective)
- Using different perspectives (rather than the person being interviewed)
- Identifying how the objectives in question (specific to the interviewee) relate to the strategic objectives
- Identifying how the objectives in question (specific to the interviewee) relate to the more generic objectives
- Structuring the objectives that have already been defined may help defining some objectives that have not yet been identified
- Quantifying objectives includes the development of a value model, and further clarifies the objectives

Keeney (1992) provides a list of nine requirements (desired properties) for the fundamental objectives, listed below:

1. *Essential*, to indicate consequences in terms of the fundamental reasons for interest in the decision situation.
2. *Controllable*, to address consequences that are influenced only by the choice of alternatives in the decision context.
3. *Complete*, to include all the fundamental aspects of the consequences of the decision alternatives.
4. *Measurable*, to define objectives precisely and to specify the degrees to which objectives can be achieved.
5. *Operational*, to render the collection of information required for an analysis reasonable considering the time and effort available
6. *Decomposable*, to allow the separate treatment of different objectives in the analysis.
7. *Nonredundant*, to avoid double counting of possible consequences.
8. *Concise*, to reduce the number of objectives needed for the analysis of a decision.
9. *Understandable*, to facilitate generation and communication of insights for guiding the decision-making process.

Keeneyís nine points, reprinted above, provide a good and concise set of generic requirements for the identification of objectives. However, this type of data gathering and analysis can be quite challenging for each specific case, and the process is not easily boiled down to a ì cookbookî or ì several easy steps.î Often, there are a number of technical, organizational, and cultural issues that can complicate this process. It is important that the facilitator has experience in conducting such studies in different environments and conditions, and the creativity to adjust to circumstances.

2.2.2 Identifying Alternatives

Once the decision maker has developed a description of the objectives, he or she must identify the alternatives that are ì pathwaysî to the objectives. Again, this information is obtained through discussions, interviews, and workshops with the stakeholders and decision makers.

Each alternative may require different resources, lead to different results, and include different levels of risk. To keep track of different alternatives, a modeling framework is needed to ì map outî the alternatives. Modeling is addressed in the section below.

2.2.3 Constructing Models to Assess Impact of Decisions

Two traditional methods for modeling decisions are influence diagrams and decision trees. Both provide a graphical representation of decision relationships and include rules for structuring the decision-making process.

Influence diagrams include the following components/elements:

- Decisions, represented by rectangles

- Chance events, represented by ovals
- Calculations or consequences, shown as rectangles with rounded corners
- Arrows between elements represent influence of one element on the other

Influence diagrams allow us to structure the decision-making process and better understand the relationships affecting the decision-making process. The best way to describe influence diagrams is on a practical example, such as the one shown in Figure 2ñ3.

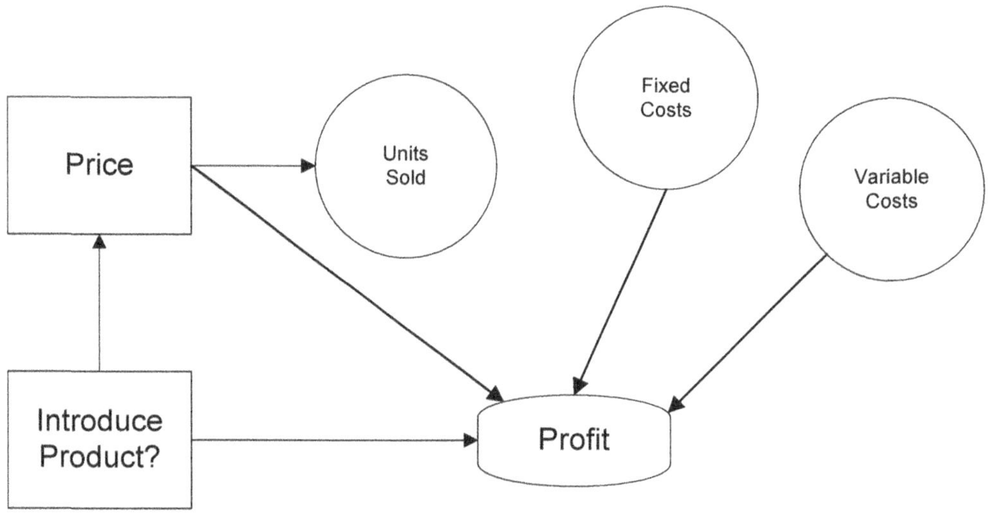

Figure 2-3. Example of Influence Diagram. From Making Hard Decisions, An Introduction to Decision Analysis 2nd ed. by Clemen. © 1996. Reprinted with permission of Brooks/Cole, a division of Thomson Learning: www.thomsonrights.com.

The example above is the influence diagram for the introduction of a new product. The fundamental objective in this decision is profit for the company. Two decisions need to be made: whether to introduce a new product, and how to price the product. Arrows indicate how different elements (referred to as ì nodesî in an influence diagram) affect each other. Please note that the decision on pricing will affect profits, as well as the number of units sold. For more examples of influence diagrams, reader is referred to Clemen (1996).

Although influence diagrams can provide a good tool for mapping out and enhancing the understanding of the decision-making process, the diagrams normally do not include enough detail to support the analysis of different alternatives. For this purpose, decision trees are used.

Decision trees use the same set of elements: decisions, chance events, and objectives. The trees ì branch outî into different alternatives, and thus visually represent different choices that are available to a decision maker. Again, it is easiest to explain the concept of a decision tree on an example, as shown in Figure 2ñ4.

The decision tree shown in Figure 2ñ4 shows two decisions. The first decision is whether to conduct research of a new product. The cost of research and development is $2 million, and the chance of obtaining a patent is estimated at 70%. The end of the line indicates the financial result of the selected path or alternative.

If the first decision is the go ahead with the research and development effort, and the patent can be obtained as result of research, the second decision needs to be made: whether to license the technology or sell it directly. The cost of developing marketing and production capacity is $10 million, and the demand can be predicted as indicated: 25% chance that the demand will be high, 55% chance that the demand will be medium, and 20% chance that the demand will be low.

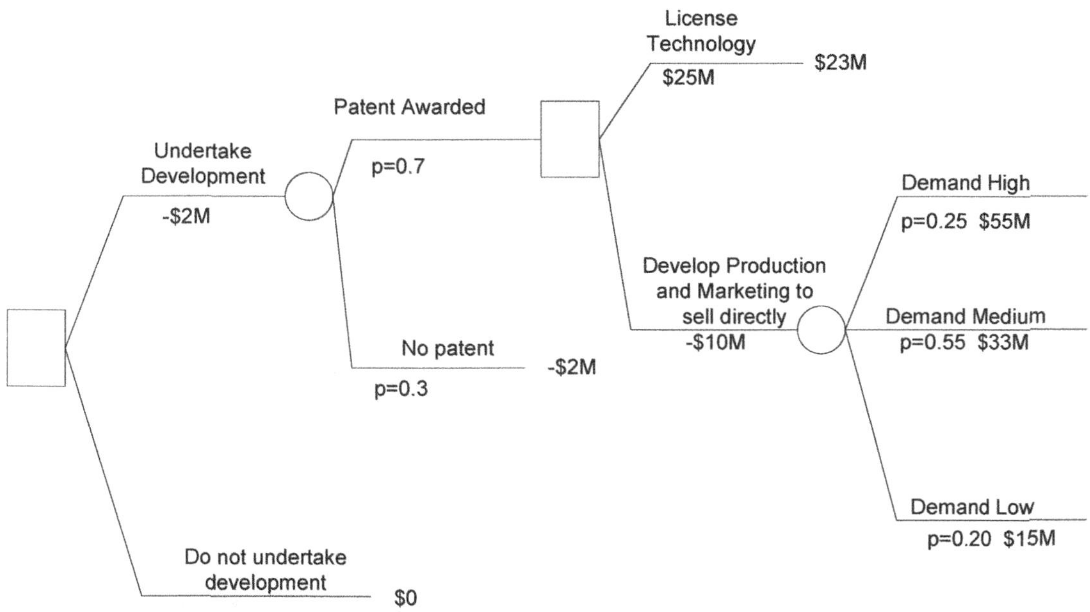

Figure 2-4. Example of Decision Tree. From Making Hard Decisions, An Introduction to Decision Analysis 2nd ed. by Clemen. © 1996. Reprinted with permission of Brooks/Cole, a division of Thomson Learning: www.thomsonrights.com.

Decisions are difficult because uncertainty makes predicting specific outcomes difficult. Nonetheless, decisions must be made, although some things beyond the control of the decision maker can have a profound impact on the outcome. Probability is the best tool for handling this uncertainty. Assigning probability to an outcome provides a mechanism to assign an apparent value to a specific choice. Further detail on how decision analysis is used for solving specific practical problems is found in Clemen (1996) and Skinner (1995).

2.3 Sensitivity Analysis

When building models of the decision-making process, it becomes clear that the outcome will depend on a number of different parameters. Often, the exact value for these parameters is not known at the time of a decision; only a feasible range and probabilities associated with each parameter may be known.

As part of the analysis process, it is important to conduct sensitivity analyses so that the decision maker can determine how ì sensitiveî the outcome is to variance in these different parameters. This helps understand the limitations of the model and makes the overall risks more predictable.

Several software packages are available today to support decision analysis, including DATA software (Tree Age Software, Inc.; www.treeage.com), Decision Programming Language (DPL; Applied Decision Analysis, Inc.; www.adainc.com), and CrystalBall risk analysis software (Decisioneering, Inc.; www.crystalball.com). These packages include tools for sensitivity analysis.

CHAPTER 3.0

HOW IS DECISION ANALYSIS RELATED TO DSS?

The preceding chapter presented a brief introduction to decision analysis. How does this theoretical discussion relate to the everyday needs of a practicing manager or engineer at a municipality? The practical lessons drawn from the discussion of decision analysis include the following:

- There exists a field of science and an applied body of work that can help us understand how to approach the decision-making process in a systematic, rigorous, and scientific way.
- Decision analysis includes methods that can be used to define objectives; this is a very important step in the development of any DSS and therefore the techniques from decision analysis are directly applicable.
- Decision analysis also provides methods for establishing a framework (model) for identifying alternatives and making choices between alternatives that are focused on the stated objectives.
- Decision analysis also provides methods and tools that may be used to represent different alternatives and evaluate the effect that each alternative would have on the objectives.
- Most decisions include uncertainty. A robust decision analysis process should include methods that consider and use probability to describe this uncertainty within the model of a decision-making process.

During the course of their work, managers and engineers routinely need to make decisions. However, not all the decisions require, or would be helped, by a decision support system. There is no definite ìlitmus testî to recognize the nature of a decision, but, in general, the types of decisions are determined based on their potential impact.

Strategic decisions are those that are broad in scope (issues and impacts across different business units) and have significant long-term consequences for the organization. Examples include selecting a site for a new treatment plant, or determining rates for users. Strategic decisions are typically ìone timeî or infrequent decisions that have major consequences.

Tactical decisions deal with business processes and issues that are mostly within the scope of one business unit and have a required frequency (how often they need to be made) that is higher than for a strategic decision. The impact of a tactical decision is thus more limited both in time and in terms of impact.

Operational decisions are often repetitive and are made at high frequency: in some cases they need to be made ìreal time.î Often the rules for these decisions are (pre)defined and the decisions are made in response to different conditions.

Each of the decision types listed above can be enhanced by a DSS. The basic approach to design an appropriate decision support, however, needs to recognize what type of a decision the system will be supporting.

CHAPTER 4.0

THE CONTEXT FOR DSS

Discussions about DSS often focus too soon on technology, and on issues related to software applications or products, but the development of a DSS is a process that must start from the *business needs*. Before declaring a decision support system to be *the* answer, or even *an* answer, users should first consider the *questions*: that is, if DSS is the solution, what is the problem? Typically, the problem will not be clearly defined, or understood, early on. The initial phases of DSS development should be focused on problem definition and formulation.

The process for identifying the requirements for a DSS can be approached from different ì anglesî including decision analysis, business process analysis, and software development. Each of these possible angles has its own tools and methods, but they also have much in common. The basic components of the problem definition aspect common to each approach are:

- Identify (fundamental and means) business objectives; this is normally done through interviews and facilitated workshops.
- Identify possible alternatives that can be evaluated; interviews, facilitated workshops, and brainstorming session may be used to define the possible approaches/alternatives.
- Develop a model of the business process or the decision-making process; the purpose of the model is to describe the business process, or decision-making process, in a rigorous way. The model provides a framework for evaluating alternatives. Different modeling tools, approaches, and notations can be used for this:
    - Modeling tools that come from decision analysis (using influence diagrams, decision trees, etc.). Such tools include software programs such as DATA and DPL. These programs allow the user to define the objectives, the decision context, and to simulate the different alternatives. Sensitivity analysis tools are included within these programs to analyze how different parameters impact the objectives. The symbolic notation used in these programs is adopted from the field of decision analysis.
    - Modeling tools for business process analysis; these tools allow the user to describe the business processes in deterministic terms and simulate different alternative scenarios. Business process descriptions within these tools include resources required to perform certain tasks, the relationship between different tasks, and also include powerful statistical analysis features.
    - Methods from software development; one of the fundamental issues in software development is defining what exactly the computer programs need to do for the user. Therefore, a number of tools were developed within Computer Aided Software Engineering (CASE) to define the underlying business

processes that the software will address. Several notations are available within such modeling including Grady-Booch notation for object-oriented systems, different formats for representing the relationship between data, entity relationship diagrams, etc.

- Using the model, evaluate alternatives (in terms of impact on the objectives) and select (and implement, if this is an automated system) the best alternative.

In some cases, the activities listed above can be applied to a significant, ì one timeî strategic decision. In such cases, the product of the entire process may be a single decision (e.g., site for a new treatment plant). In other cases, it may be a matter of providing an ì automatedî tactical or operational solution that is executed frequently and may include significant automation and limited human input. Under the latter scenario, the DSS may be a computer program providing assistance to engineers or operators who are trying to deal with routine business issues such as process control, maintenance management, or security issues, or to managers who are preparing the budget for the organization.

Several methods are used to determine the business needs of organizations and analyze their decision-making processes. Examples and case studies of how such methods are employed in the municipal water/wastewater sector are provided in Appendix C.

CHAPTER 5.0

METHODS FOR IDENTIFYING KEY BUSINESS ISSUES

Because decisions are the results of business processes, the development of a DSS requires that we first understand the underlying business processes and issues. This section discusses the methods that can be used to determine the characteristics of the business processes that would be supported by the DSS.

The method used on this project was based on use cases; this approach has been applied to the development of user/business requirements for a range of different computer systems. According to Armour and Miller (2000), the use cases approach allows decision makers to

- capture requirements;
- envision how applications would work;
- delimit the system boundaries; and
- communicate with system developers and users.

Use cases provide a dynamic view of the system, a basis for user interface design, and a base for allocating functionality to components. Use cases allow users to define object interaction, data requirements and access, and testing and integration needs. Use cases can become a standard way of representing business processes for an organization or business.

5.1 Tools for Defining Business Needs

The main vehicle/mechanism for defining use cases on this project was an application that was accessible to all project team members via the Internet. This application was developed for the specific needs of this project. The requirements for the tool were as follows:

- Information about the use cases would be stored in a database to allow for better analysis of the results.
- All project team members should be able to work on use cases simultaneously from different geographic locations, and need to see (online) the changes made by the other team members.
- The use case tool/application needs to capture both the characteristics of each business process and the overall hierarchy of business processes
- The tool needs to capture the data that is required to conduct each business process and organize the entries accordingly.

The tool for defining use cases was a Web browser-based front end to a Microsoft Access database. The tool was placed on a server connected to the Internet and made available to all team members. Each project team member was provided with a user ID and a password that allows for immediate, concurrent, and interactive access to the database.

When the uniform resource locater (URL) was accessed, the screen shown in Figure 5ñ1 requested the user ID and password.

Each team member was provided with his own access ID and password, and this allowed the program to track the identity of the person who makes the changes to the database.

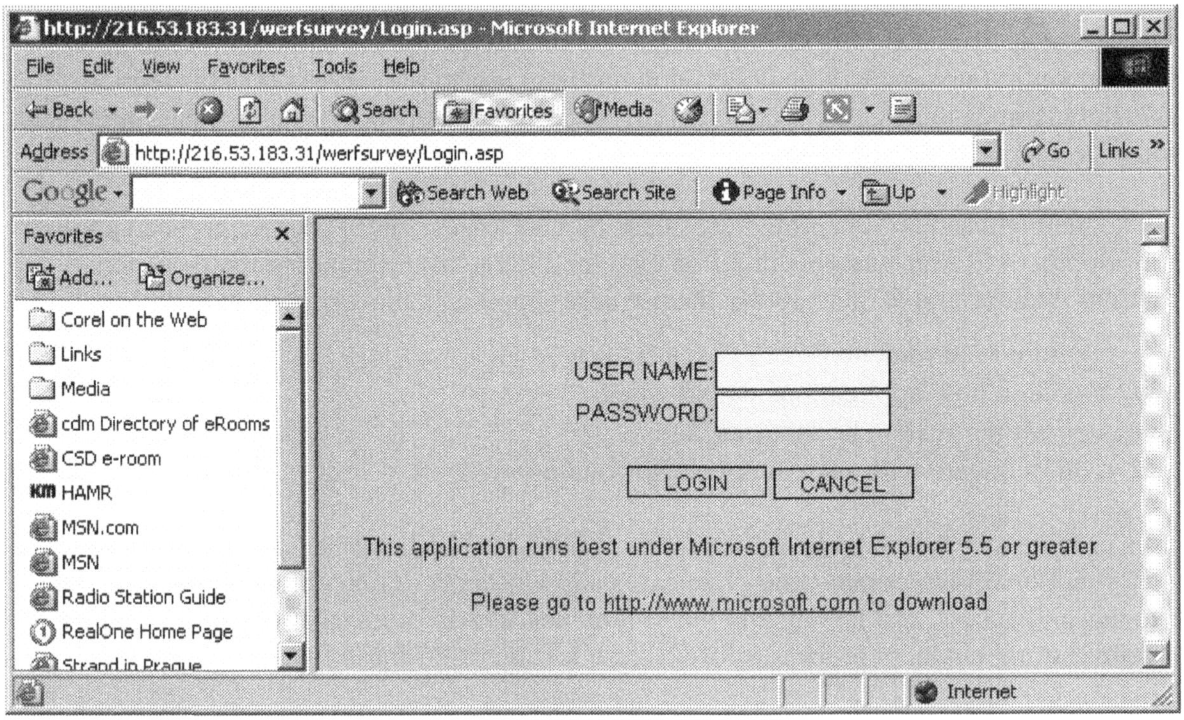

Figure 5-1. Use Case Web-based Application Login Screen.

Once logged in, the screen in Figure 5-2 appeared.

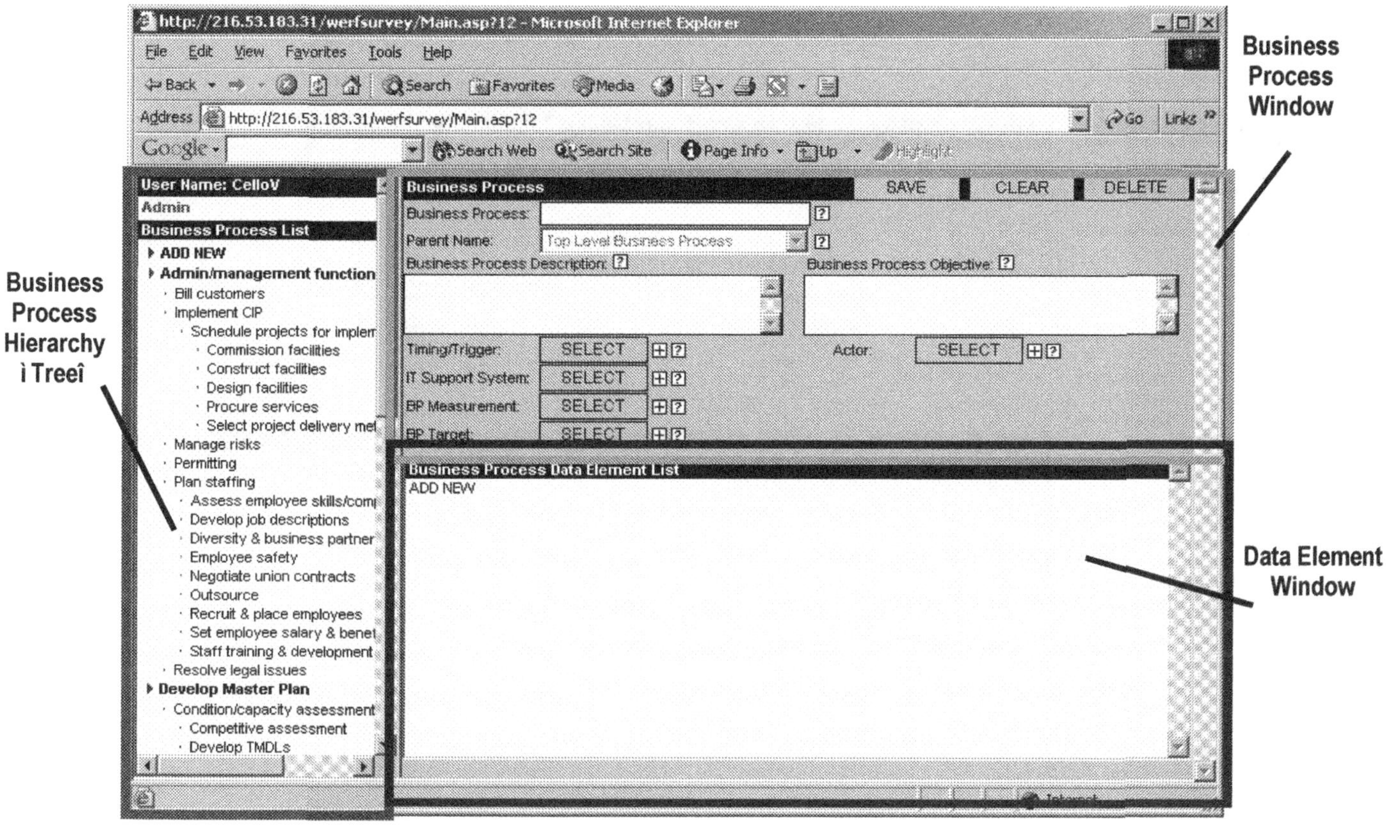

Figure 5-2. Use Case Web-based Application Main Screen.

The window shown in Figure 5-2 includes three regions: the business process hierarchy (ì treeî), the business process window, and the data element window. The roles for these regions/components are as follows:

1. The ì Treeî allows the user to select an existing business process or to add a new business process, thus navigating and modifying the overall ì mapî of business processes
2. The Business Process window allows the user to define the characteristics of the use case for this business process by entering information regarding the different elements of the use case. Please note that when the system first comes up, the pointer will be on the top business process in the hierarchy. To select another business process, the user needs to click the left mouse button on the Tree.
3. The Data Element window allows the user to define the information that is used within this business process, and thus the information that is required to make a decision that will be the result of the business process. A number of different data elements may be defined (Figure 5-3). Each of the data elements is described with a block of information that is associated with it.

When a specific Business Process is selected, the Data Element window will show the list of data elements needed in support of this business process (Figure 5-3).

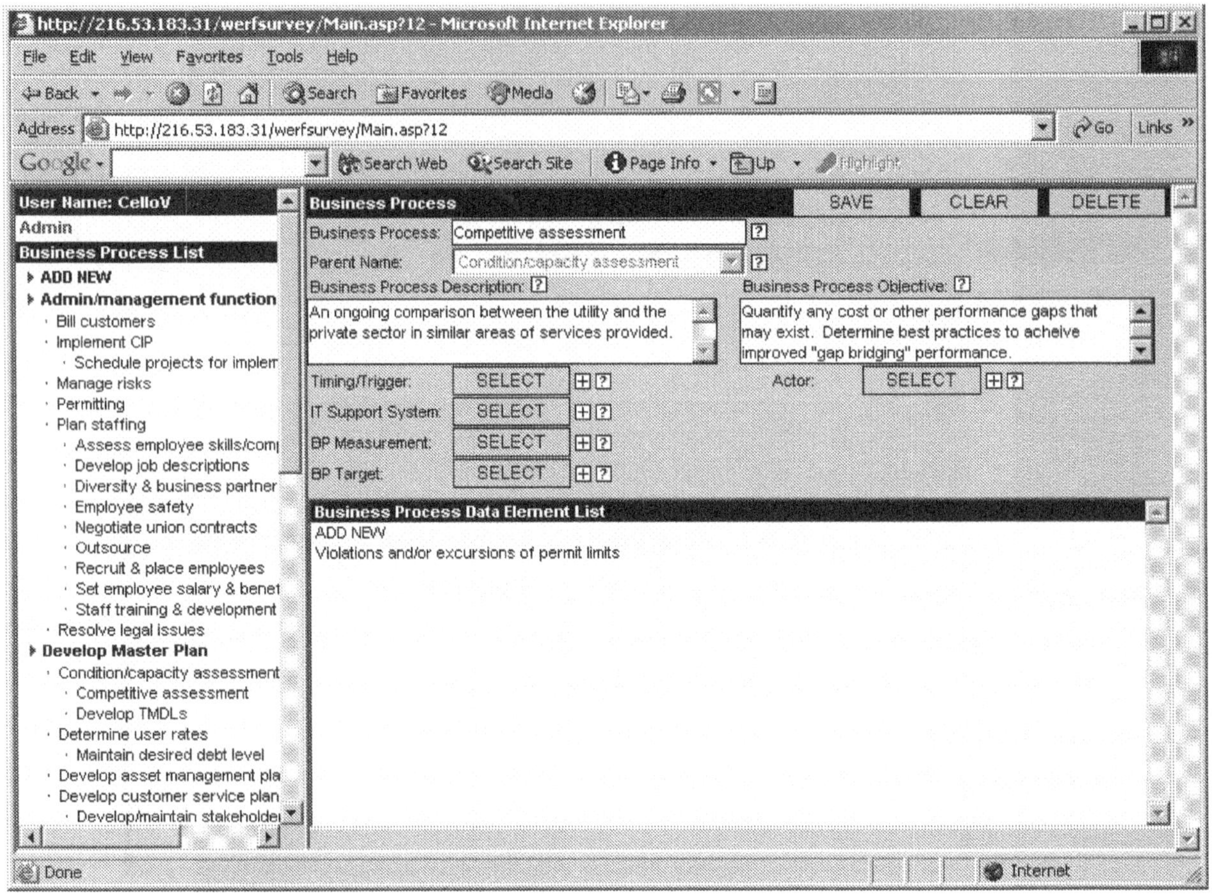

Figure 5-3. Use Case Web-based Application Data Element Listing Screen.

When the user clicks on a specific item in the list of data elements (the example here is ì Violations and/or excursions of permit limitsî), a window containing the definition of this data element will appear, as shown in Figure 5-4.

This use case tool included a Web-based entry form to a server-based Access database. All user entries were immediately updating the database on the server. This allowed several team members to work on the same database concurrently, to view the input that others have provided, and to collaborate more easily.

The database that was developed through this effort provided the basis for the analysis of different business processes, and it also provided a template for a method that a municipality could use to map out its own data sources and business processes. (Note: The tool was removed from the Internet after the conclusion of the project.)

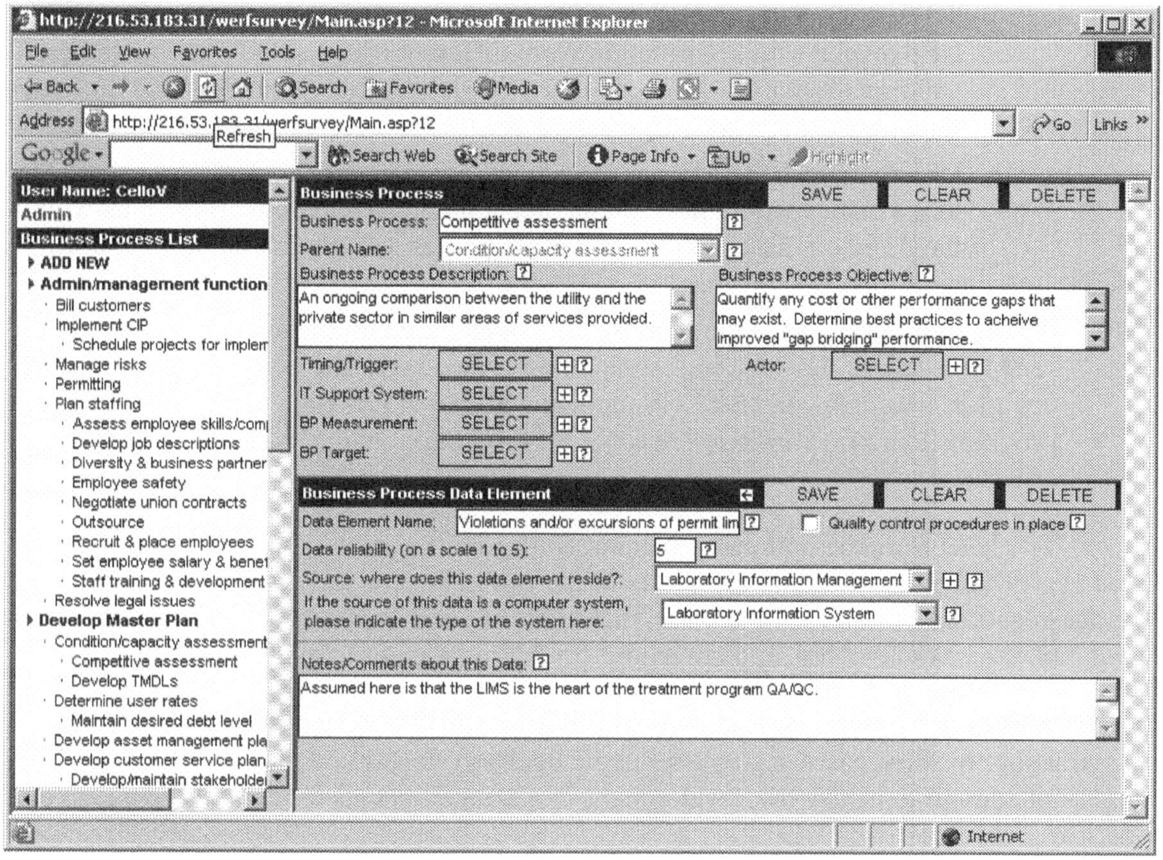

Figure 5-4. Use Case Web-based Application Data Element Detail Screen.

5.2 DSS Can Play Different Roles

A DSS can function at different levels, and play different roles in the overall decision-making process. One way to differentiate between the different ì flavorsî of DSS is to consider the role that the DSS will play with respect to a human in the decision-making process. The range of options includes the following:

1. Support: DSS acquires and processes information, and presents it to the decision maker. The decision is still made by a person, and the DSS provides (some or all of the) support functions as follows:
    a. DSS collects and organizes information from different sources that may include different computer systemsó including Supervisory Control and Data Acquisition (SCADA), Laboratory Information Management System (LIMS), Computerized Maintenance Management Software (CMMS), Geographic Information System (GIS), modelsó and databases.
    b. DSS processes ì rawî data to calculate new, more complex computed variables that provide the decision maker with better insight and understanding of the process state (e.g., dissolved oxygen uptake profile in an activated sludge plant,

CIP-wide capital project expenditures, forecasted (future) inflows to the treatment plant)

c. DSS displays the information from different sources and systems in an effective and user-friendly way, so that the decision maker can understand the process effectively

d. DSS evaluates the outcome of several preselected alternatives to show the impact that a choice would have on the decision goals and objectives. This would allow the decision maker to play ìwhat ifî games and predict the impact of different decisions.

2. Automation: DSS can actually make decisions based on the available data and pre-established rules. Such an implementation may not require active interaction with the operator. Examples of such DSS include automated real time control systems, or programs that prioritize work tasks (e.g., inspection, maintenance) based on predefined criteria.

To chose from the options above and configure the most appropriate approach, it is necessary to carefully examine the nature of the problem. The factors to be considered include the organizational structure and culture, the reliability and accuracy of the information that is available, the risks involved, and the nature (including the complexity) of the problem.

A DSS can provide ongoing support to many different professionals across the organization. In some cases, a single strategic decision (locating a major facility) may be important enough to justify the development of a DSS that is specifically targeted for that decision.

CHAPTER 6.0

IT ASPECT OF DSS IN WASTEWATER AGENCIES

At a typical large wastewater utility, a number of different computer systems have been purchased, configured, and implemented during the past 20 years to handle specific and separate business issues within the organization. This progress has not been uniform: some agencies have implemented automation in many areas, whereas others have implemented only a few systems. DSS would try to address limited access (reporting and viewing) and restricted analysis. Individual, primary computer applications ì ownî different types of information related to the management of a wastewater utility. Access to this information is restricted: in most of the cases data can only be retrieved through that individual application. Therefore, reporting and viewing is limited. Analysis is also limited to the primary applications, and therefore restricted to a single technical domain.

The primary applications create and own the data. The DSS collects, integrates, presents, and analyzes enterprise data (from primary applications) but does not own or create any original data. The DSS analytical tools may create derived data. The relationship between the DSS and the primary applications is shown in Figure 6-1.

As shown in Figure 6-1, a DSS would include software that connects DSS to other systems to access the data from these systems and make it available to DSS. In addition, the decision support database may have pointers to information that resides in other systems, and in some cases duplicates of data from other systems

The DSS may include configuration tools that perform the following functions:

- Configure the interface between the primary applications and the DSS (labeled ì 1î in Figure 6-1)
- Organize the information in the DSS database (2)
- Configure the presentation (user interface) of information to users (3)
- Configure the interface between the DSS and the analytical engines and agents that operate on the integrated, synthesized data (4)

DSS analytical modules may include ì agentsî and external computational ì engines.î Agents are computerized procedures that operate on the DSS database and generate derived data: for example, a ì detection agentî may search through the DSS database for a predefined pattern (a set of conditions) and generate an alarm related to security or operational issues. ì External enginesî are applications that may interact with a DSS: for example, a hydraulic model may be used by the DSS as an external engine to compute the hydraulic profile between the two points where levels are measured in the sewer network

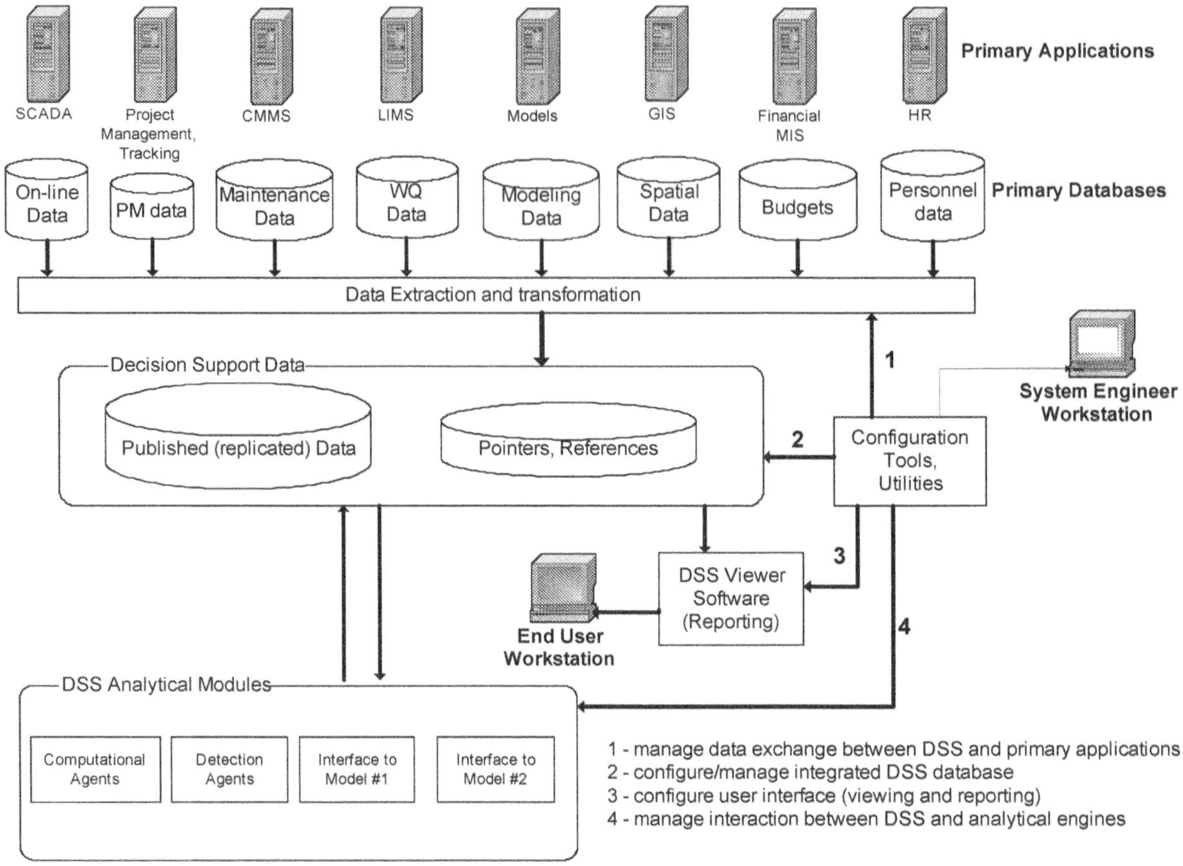

Figure 6-1. Context for DSS: Generic View of Components Within Overall Architecture.

The diagram presented in Figure 6-1 is generic and conceptual; implementation of these concepts could be different, using different methods and tools. It is also possible to envision and develop a more limited version of DSS that might not include all of the components shown in Figure 6-1. The primary applications are focused on providing as much power and flexibility as possible to users within a specific business domain. Use of these primary software tools normally requires ìdomain expertiseî on a fairly detailed level, and also specific (often extensive) training in the primary software applications (such as GIS, SCADA, models, CMMS, and LIMS). The emphasis for DSS is on synthesis of diverse information from multiple sources so that decision makers can extract more meaning and content from the data.

6.1 DSS Technology

The main DSS business goals relate to three aspects of technology (software) that would be used to deliver the solutions:

- Integration: DSS collects and organizes the information that is created by the primary applications and is then ìpublishedî to the DSS database. The DSS makes this information available for access by anybody with proper access privileges.

- Viewing/reporting: once the system wide data from different sources/applications is available, a set of tools within the DSS provides a mechanism for accessing and combining the information for presentation and reporting.
- Analysis: the primary applications often include specialized analytical tools that analyze the information that they collect. The DSS opens up new possibilities for processing the integrated information and adding analytical capabilities on top of the integrated data. Therefore, the scope of analysis can be broader because the analysis is based on enterprise-wide data.

A DSS may address the business needs of users on one of several different levels in an organization: operational, tactical, and strategic. Ideally, the same DSS should be able to extend into all three areas. A critical aspect of such a broad DSS is to be flexible and retain the ability to meet different needs for information from different users and on different levels. Flexibility to address diverse needs could be achieved by reusing the same standard components.

DSS can meet these goals by leveraging the investments made into the different computer applications and the wealth of data that may be ìhiddenî on individual computer systems. The information that DSS makes available is already available: it has already been paid for.

6.2 Technology Options

The need for integration of computing systems, and for decision support, has been receiving attention in many areas of private enterprise. Major industrial sectors and large companies have been leading efforts to develop decision support systems.

It has been especially difficult to prepare an overview of the available technology for system integration because software products in this area are undergoing rapid development and change. There is also considerable risk that the status of the marketplace described today may not be accurate or relevant in a few months. The ability to provide integrated systems is viewed as a high priority by the software vendors, therefore this area is extremely competitive and volatile.

In the 1980s it was relatively easy to get unbiased assessments of software products. Today, the number of available products has increased dramatically but, in many cases, the information that is available about the performance of these products comes directly from the vendors and lacks objectivity. Products are described in glowing terms and often a similar set of terms (that happen to be popular at the time) describe different products. According to the information that is easily available, most of these systems are object oriented, powerful, open, reliable, secure, and above all, very user friendly. It is very difficult to obtain an independent verification for many of these claims.

The choices of off-the-shelf technology for integration seem to have the following general options:

Single vendor. The user may try and solve most business problems within the suite of products provided by a single vendor. Some large software vendorsó such as IBM, SAP, PeopleSoft, or Oracleó provide applications that address a wide array of business-related functions. They are also keenly aware of the need for integration and they promote integration within their suite of products. IBM for example, offers a suite of products that constitute an ìintegration platformî under the name Websphere; SAPís integration platform is NetWeaver, and BEA provides a system called WebLogic.

Software application links. A number of emerging companies provide software that link different software applications; they provide modules called application programming interface that can be used to integrate different applications, often from different vendors. Such companies include iWay, Attunity, and Taviz.

Information exchange. Several products provide an equivalent to an ìinformation busî that uses an equivalent of a messaging system to exchange information between different systems. An example of such technology is software from Tibco.

Middleware. Software products that help connect other applications are sometimes called ìmiddleware.î The battle between middleware makers and main application developers for this lucrative market is fierce[1] with an uncertain outcome.

It is important to recognize that there is a significant difference between an off-the-shelf program and having a fully integrated system. Although some of this software has fairly high price tags, the cost of actually implementing integration can be considerably higher than the purchase price of the software package.

An additional issue is the nature of the wastewater utility and the business processes that are involved in managing a wastewater agency. Municipalities not only have a need for purely transactional systems (accounting, personnel, billing, etc.), but they also need to integrate the sophisticated analytical tools that are used to address the scientific or engineering issues that are very important for the performance of wastewater utility. Most commercially available systems are structured to function within the transactional paradigm and may be inadequate for full range of business processes encountered in a wastewater utility.

The diversity and complexity of applications highlight the need to define the specific functional requirements for DSS that would work well for a wastewater agency.

6.3 Integration of IT in Wastewater Agencies

Many public sector agencies have realized the need to integrate their computer systems, and they are taking different approaches toward integration. The business need for integration is often initially realized when it becomes apparent that two computer applications could benefit by sharing data. A decision can then be made to connect two applications directly. Examples of business reasons behind such direct (point-to-point) connections between two different computer systems are as follows:

---

[1] From InfoWorld, June 20, 2003, SAP advances on Websphere turf by Ed Scannell, Mark Jones: ìCraig Conway, CEO of Pleasanton Calif.-based PeopleSoft, used his CeBit keynote last week to fire a salvo at integration players. ìThis is the beginning of the end of middleware. Itís the beginning of cross-functional applications, where the integration between them is not a burden to you, itís a burden to the software companies.î

1. GIS (maps) and mathematical models for the sewer system: it is much easier to view and analyze sewer models, for example, if they can be presented within the context of a map. Therefore, connecting the models to the GIS makes a lot of sense.
2. CMMS and SCADA: some of the information that is created and maintained within the CMMS (such as status of work orders) can have value for operators: it would be nice for the operator to know the status of a work order that was issued for a piece of operational equipment. On the other hand, online information about run times for pumps is relevant and useful for CMMS.
3. SCADA and weather forecasting: operators can clearly benefit from advanced knowledge of weather
4. SCADA and facility inspections: operators (main users of SCADA) would be interested in knowing the status of the inspections
5. Mathematical models and SCADA: models can utilize SCADA-based input to create more meaningful measurements of process performance (e.g., dissolved oxygen uptake rate, hydraulic profiles, etc.)
6. Customer information system and inspections: for the staff person answering customer phone calls about problems such as flooding or odor, it may be useful to have access to information about inspections of facilities.
7. Project management system and budgets: information about the capital project budgets and the project expenditures
8. Mathematical models and inspections: if, for example, somebody is modeling a sewer network, it would be good to know if some parts of the network have significant sediment deposits (and this information normally resides in the Inspection database)

Although each of these cases has a separate and valid business reason for integration between two systems may be identified, the individual connections between different systems create a complicated web of dependencies that can become unmanageable. Therefore, point-to-point integration may initially appear attractive because it seems like the problem is being addressed in a gradual, phased manner. However, over time these individual connections between systems can become a significant software maintenance problem.

CHAPTER 7.0

INTEGRATION AND DSS IN PARALLEL (PRIVATE) INDUSTRY

An important goal of this project was to summarize the efforts related to DSS in parallel industries: those private (for-profit) industries that have some similarities with the wastewater industry (which is mostly in the public sector). These parallel industries have invested heavily into DSS, and there was thus much to report on.

Because the volume of information on DSS applications in parallel industries is substantial, even a summary turned out to be large. Therefore, this summary is provided as Appendix A of this report: this way, it will be easily accessible but still not overwhelm the main body of the report.

CHAPTER 8.0

INTEGRATION AND DSS IN WASTEWATER INDUSTRY

The development and implementation of DSS is still fairly new within wastewater industry. Although many have recognized the need, the approach is far from standard or uniform, best practices have not been defined, and there is still considerable diversity in efforts to integrate the computer systems and develop DSS.

Different companies providing services related to integration of DSS in the wastewater industry approach these issues from different perspectives and offer different solutions. Several leading companies (including CH2MHill, Malcolm Pirnie, Westin Engineering, EMA, Inc., CDM, and Black & Veatch) that provide related services in this area were contacted and asked to contribute a description of their approach to system integration and decision support. Several companies agreed to submit material that describes their approach and practice in these areas, and their contributions are included in Appendix C.

In some cases, internal agency staff takes the leading role in both planning and implementing system integration. Examples of integration projects that were led and executed ì in houseî include Massachusetts Water Resources Authority in Boston and Clean Water Services in Oregon.

CHAPTER 9.0

FUNCTIONAL USER REQUIREMENTS

The functional requirements specified in this document are generic. These requirements do not include the level of detail often required by a developer to start programming, but provide a general and generic functional framework for a DSS.

The software industry has struggled to find a good way to describe what a software program should do. Humphrey (1990) writes:

> While we sometimes get firm requirements, a software perversity law seems to dictate that the firmer the specifications are, the more likely they are to be wrong. The fact is that the demand for firm and unchanging requirements is mostly wishful thinking and may even be an unconscious attempt to avoid blame for a poor resultÖ .With rare exceptions, the requirements must change as the software job progresses. Just by writing a program, we change our perception of the task. The ripple of change starts with the requirements and continues through implementation, initial use, and even through later enhancements. Even if we could plan for all this development learning, the computer implementation of human tasks changes the applications themselves. Requirements by their very nature cannot be firm because we cannot anticipate the ways the tasks will change when they are automated.

These days, it is rare for developers to create or use large documents that include great amount of detail regarding the functionality of a software system. Systems are developed in an evolutionary, iterative way and the level of detail is normally kept at a fairly coarse level. As a minimum, however, system requirements should address the following:

- Who will use the system? Different (groups of) users are described.
- What functions should be available to (different groups of) users?
- What will the user interface look like?
- What type of information will be handled by the system?
- How will this system interact with other computer systems?
- How will security issues be handled?

The main purpose for the functional requirements is to facilitate the communication between the users and the engineers and software developers. The functional requirements should therefore meet the following criteria:

- Clarity: are the ideas and concepts communicated clearly?
- Feasibility: can the scope of requirements be done with the available budget?

- Completeness: do users know what will be delivered to them?

This report will not provide a detailed requirements document. The scope of this project was not to provide full detail of information sufficient for a software developer to start coding the system, but rather to propose a viable (but still conceptual) framework for developing a DSS. The ideas and concepts presented in this report therefore relate to generic needs.

CHAPTER 10.0

DATA REQUIREMENTS FOR DECISION SUPPORT IN A WASTEWATER AGENCY

Wastewater agencies have made significant investments into their IT infrastructure. For example, according to a recent study (King County, 2002), the wastewater division of King County, Washington, invested $42 million into IT from 1992 and 2002. The same study recommends a level of investment for nine major IT projects for the next decade at $46 million.

One of the driving forces for this project was a sense on the part of managers in many wastewater agencies that the amount of data that is being gathered through a multitude of installed computer systems and applications is overwhelming. Managers would like to make sense of these data and identify which data are really needed to manage a wastewater agency.

There is also a concern that the large volume of collected data makes quality control difficult. The managers believe that it would be helpful to generate a parsimonious list of data that is required for decision support.

Identifying data needed for decision support is very important, but this research did not (and could not) specify the ìultimateî data list that should be used in every agency. Instead, this research recommends a process for identifying the required data. The process for identifying the required data for decision support starts with the identification of the business processes, and the specification of the decisions that are part of each business process.

A tool for defining business processes, as briefly summarized in the example presented earlier in this report, includes functionality to specify the data that are required for making decisions related to each business process. Therefore, using this approach and this method, users can identify the data required for decision support by first defining the business processes and the required decisions needed. A hierarchy of business processes was also defined within this tool by a group of four wastewater agency managers/professionals, to provide the user with a guide and a starting point.

The advantage of this approach is that the identification of the data required for decision support is directly based on the business processes and the definition of the decisions themselves. Each municipality has its own business processes, and although there are many similarities between them, there are also significant differences. Data that are critical for one agency may have limited value for another agency.

The other reason why this approach is more practical than defining an ultimate data list is because a DSS may be used to support different decision-making processes on different levels.

Data that are important on an operational level may not be critical on the strategic level. For example, the decision to identify the site for locating a treatment plant is different and requires entirely different data from the decision on the appropriate sludge wasting rate once the plant is operational.

Although an ìomnipotentî DSS would be nice, it is far more likely that the application of DSS will start with a more limited scope and focus initially on a specific set of business processes. The scope can then be increased in a phased manner. Therefore, the ultimate data list will gradually emerge from within the organization, but will always be based on definitions of business processes.

When developing a DSS, however, it is important to recognize that systems on strategic and tactical levels will ultimately benefit from data that are gathered by systems on operational level. The information will need to propagate upward. Ultimately there should also be a feedback link downward from the strategic goals into operations.

10.1 Who Would Use the DSS Technology?

As shown in Figure 6-1, two different groups may interact with a DSS:

- System engineers are experienced users who have a high level of technical proficiency and training. System engineers set up the system, configure the integration tools and the databases, and construct the views for end users.
- End users are people who use the DSS but do not have the desire, need, or the required skills or training to do the system setup, configuration, or administration.

The focus of activities changes with the DSS development and implementation cycle: initially, there is a heavy involvement by the system engineers to establish the tools and the databases. Once the system is set up, the activities of system engineers decrease. Most of the activities are on the production level and executed by end users.

10.2 What Functions Should Be Available to System Engineers?

The system engineer is defined here as a person who is familiar with the DSS software functionality and is also knowledgeable about the databases, applications that are used by the agency, and the general programming concepts. In some cases, such a person could be employed by the agency. In many cases, consultants who have helped the agency with previous automation implementation projects may also have such skills and experience.

The DSS should be able to provide the system engineer with the following functions:

- *Application registration tools* are used to ìregisterî primary applications with the DSS. The registration process establishes all aspects of the interaction between each primary application and the DSS. These tools are conceptually represented on the overall system diagram (Figure 6-1) as the line (labeled ì1î) between the box labeled ìConfiguration tools, utilitiesî and the box ìData extraction and transformation.î For each of the primary applications, a software ìdriverî needs to be developed and configured to handle the interactions between the primary application and the enterprise-wide database.

- *Publishing tools* to configure publishing of information from the primary application and the DSS. Each individual primary software application maintains control over its own dataó in some cases vast amounts of detailed information that is unlikely to be of interest to a broader audience in the agency. Publishing tools allow the system engineer to select the information that the primary application shares (makes available) to the DSS, and configure the data exchange (frequency, units, range, etc.). On the overall system diagram (Figure 6-1), these tools are also included in the line (labeled ì 1î) between the box labeled ì Configuration tools, utilitiesî and the box ì Data extraction and transformation.î
- *Data organization tools* are used by the system engineer to relate information from the primary application to the other information that is handled by the DSS. To be useful, the integrated information must be properly organized. For example, many pieces of data that reside in different primary software applications may be related to a single physical facility (such as pipe or manhole). The DSS system must be structured to understand how these different pieces of information are related to each other. These tools are conceptually represented in Figure 6-1 as the line (labeled ì 2î) between the box labeled ì Configuration tools, utilitiesî and the box ì DSS data.î
- *Presentation configuration tools* help the system engineer (advanced user) set up the viewing mechanisms (screens or dashboards, and reports) for the end user. The system engineer uses the presentation configuration tools to prepare the user-specific ì dashboardsî (or Web pages) for different end users. These tools are conceptually represented in Figure 6-1 as the line (labeled ì 3î) between the box labeled ì Configuration tools, utilitiesî and the box ì DSS viewer software, reports.î
- *Analysis configuration tools* are software utilities that are conceptually represented on Figure 6-1 as the line (labeled ì 4î) between the box labeled ì Configuration tools, utilitiesî and the box ì DSS analytical modules.î Most of the primary applications have their own, often extensive and complex, analytical capabilities within the specific business domain that they address. The analyses performed by the primary applications require from the user a high degree of familiarity with the primary application and also expertise in the specific functional domain (e.g., process control knowledge for the user of a SCADA supervisory control system, or an understanding of hydraulics for the user of a mathematical model for the sewers). The analytical tools that should be available within a DSS include computational agents, pattern detection agents, and interfaces with different computational tools from primary applications such as mathematical models or GIS.
- *System management and administration tools* manage security and access from different user accounts.

The scope and volume of work required from systems engineers will vary during the different phases of system implementation. The highest level of effort should be expected during the early stages of system definition and initial configuration.

10.3 What Functions Should Be Available to End Users?

Often there is a conflict between making applications more powerful and flexible versus making them very safe or very easy to use. Similar issues have existed in the area of consumer products as well: for example, this is why some people prefer simple click-and-shoot cameras and others require much more control over the process of taking photographs. Although there will certainly be some more advanced end users who could easily use some of the systems engineersí tools, DSS should also serve the needs of people who want to emphasize the requirements for a system that is simple and easy to use.

End user is defined here as a person who may choose to focus primarily on the viewing and reporting aspects of DSS. The functionality for an end user who does not want to have the maximum flexibility but wants to emphasize the ease of use should include the following:

- Dashboards may be used to display the DSS information. Dashboards should include presentation components such as graphs, charts, maps, drawings, links to reports and documents, and different media (photos, video).
- Mechanisms to retrieve information from the enterprise-wide database through the DSS and publish this information to different applications (such as MS Word, MS Excel, databases, etc.).
- Prepare and print reports using the information available within the DSS.

The DSS should include tools to configure many different types of viewers that can present the available enterprise-wide information to the user.

10.4 What Could a Graphical User Interface Look Like?

During the preparation of this report, some competing objectives and priorities emerged: if this report remained only on a theoretical/conceptual level, it might leave too much of the important content up to the readersí imagination. Comments from the Project Subcommittee during the project have consistently demanded a greater level of specificity and detail. If the graphical user interface (GUI) was only described in terms of the generic and broadly used (and a bit worn) epithets such as ìuser friendlyî or ìWeb basedî, the reader might have a difficult time imagining what a DSS would look like on a practical level. On the other hand, detailed design of the user interface is not within the scope of this project.

To provide useful examples of GUI functionality, but also avoid presenting a restricted view, this research strives to reach a compromise: some examples of possible GUI functionality are shown but the reader is asked to consider this as only one of the possible solutions. Additionally, some example case studies of integration tools and DSS related projects within the wastewater industry are included in the appendices.

Considering the industry trends and the direction taken by most application developers, the optimal viewing mechanism seems to be a Web browser. The look and feel of a Web browser is by far the most familiar interface to majority of potential users.

The appendices to this report include descriptions of how leading companies in wastewater arena address the DSS. Some of their contributions include descriptions of the systems that they provide. The following sections will present some generic ideas regarding the functionality for two types of GUIs: the one that is primarily used by systems engineers to configure the system, and a GUI that would be used by a typical end user.

10.4.1 GUI for System Engineers

System engineers need to configure the system and prepare the dashboards for end users. An example of a tool that would allow system engineers to do this is shown in Figure 10-1.

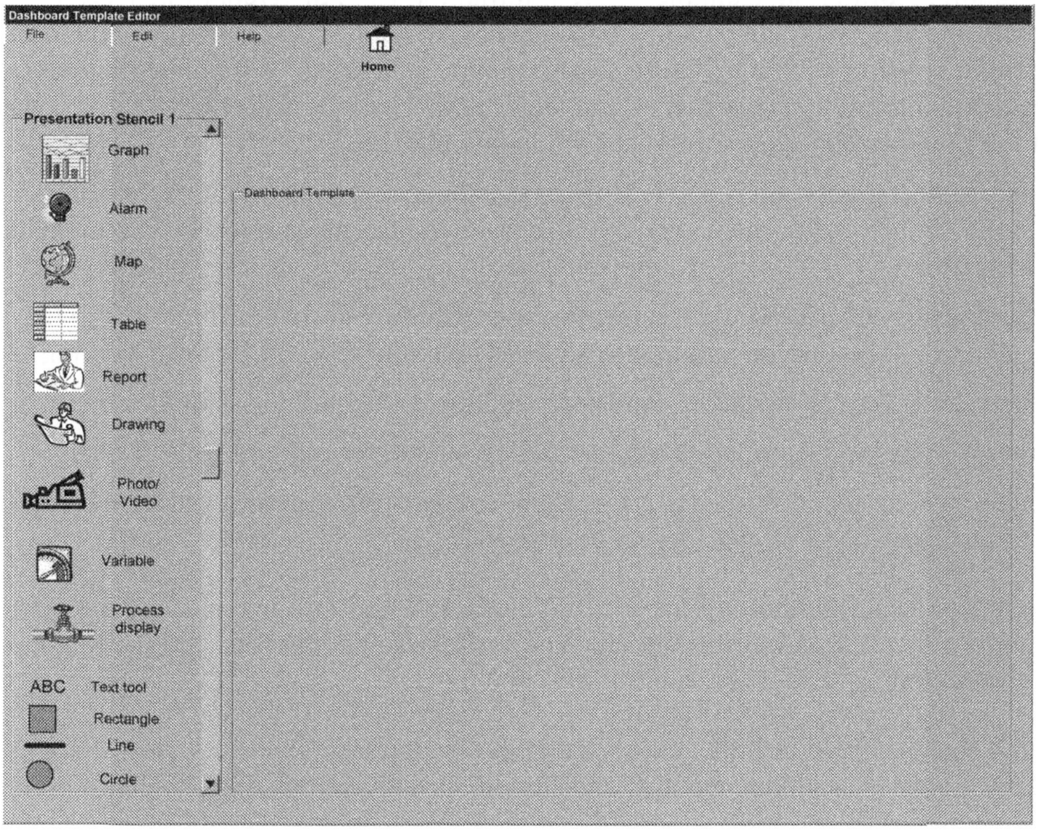

Figure 10-1. Example of Dashboard Builder for Systems Engineers.

This example shows a tool that includes different presentation components (ì stencilsî) that can be placed on the dashboard template and configured. For example, the system engineer could drag the ì Graphî object on to the template, stretch it to a desired size, and then either link it to a predefined graph or configure it using a configuration ì wizard.î

Other presentation objects would be placed on the template and configured in similar ways. The viewers for these objects depend on the data being displayed: maps, drawings, video, etc. Existing viewers are available on the market for most data formats. So, for example, drawings could be viewed through a viewer provided by a CAD vendor such as Autodesk.

10.4.2 GUI for End Users

End users can interact with the systems through dashboards that are configured by the system engineer. Case studies provided in Appendix C show how Web pages were configured for a number of different, specific implementations and projects.

The user interface provides a view into the system-wide (enterprise) data. Dashboards are configured using standard presentation components, such as:

- Graphs and charts (2- and 3-dimensional, pie charts, bar charts, etc.) that connect the information from the DSS database to the graphical presentation objects for presentation.
- Hydraulic profile is an example where information from different systems can be used to create useful derived information. Although SCADA systems normally show the status of a single measurement, a hydraulic profile can be generated from online measurements of level and information (from SCADA), configuration of the sewer (GIS), and computation of the flows and levels in a segment (hydraulic model).
- Links to documents and reports. Most reports today are prepared in the electronic format, so DSS simply makes them available. Links to documents and reports are included in the dashboard.
- Drawings can be accessed in two ways: by embedding drawings directly into the dashboard, or by providing a link to a viewer that opens up the desired drawing.
- Maps can also be provided as embedded into the dashboard, or as links to map viewers.
- Process views and summaries can be made available from the SCADA system.
- Radar/weather forecast images are available either from the Internet (at no charge) or through specialized weather forecasting service.

Design and configuration of the specific user interfaces for the end users is a process that is significantly enhanced when users are engaged early on in the process, and their input is carefully considered.

Each person should be able to view the DSS through a personalized view. Examples of some dashboards on operational, tactical, and strategic levels are provided below in Figures 10-2 through 10-4.

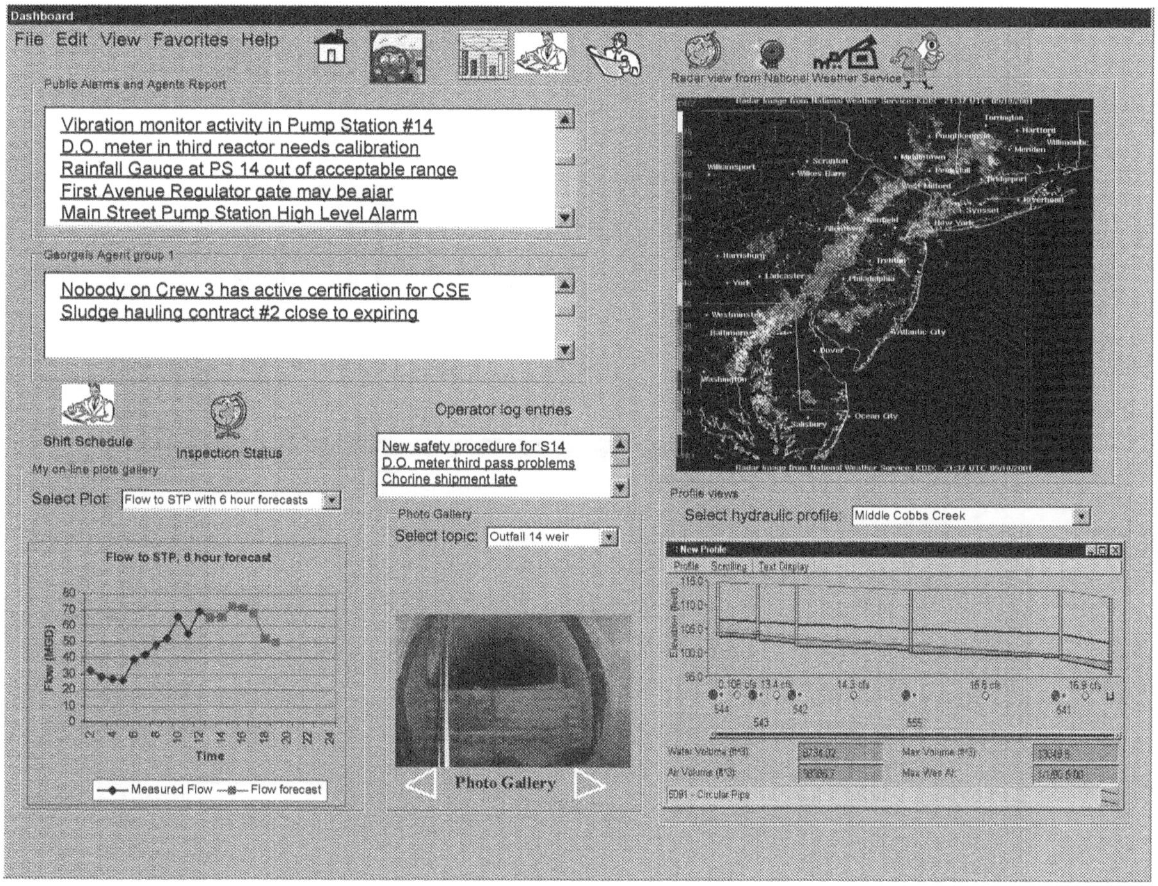

Figure 10-2. Example of an Operational End-user ìDashboardî (Web Page).

The example view on the ìoperationalî level shown above in Figure 10-2 includes views into the information that comes from the maintenance management system (CMMS), weather forecasting information available on the Internet, GIS, mathematical model, SCADA, and a facility manual. Some of the information above is online: flow to the treatment plant, from SCADA, and the hydraulic grade line in the hydraulic profile (based on SCADA readings and a model interpolation, and GIS info on the pipe).

This example page also includes information provided by the DSS ìagentsî including forecasted flows into the treatment plant, and status of contracts or certifications.

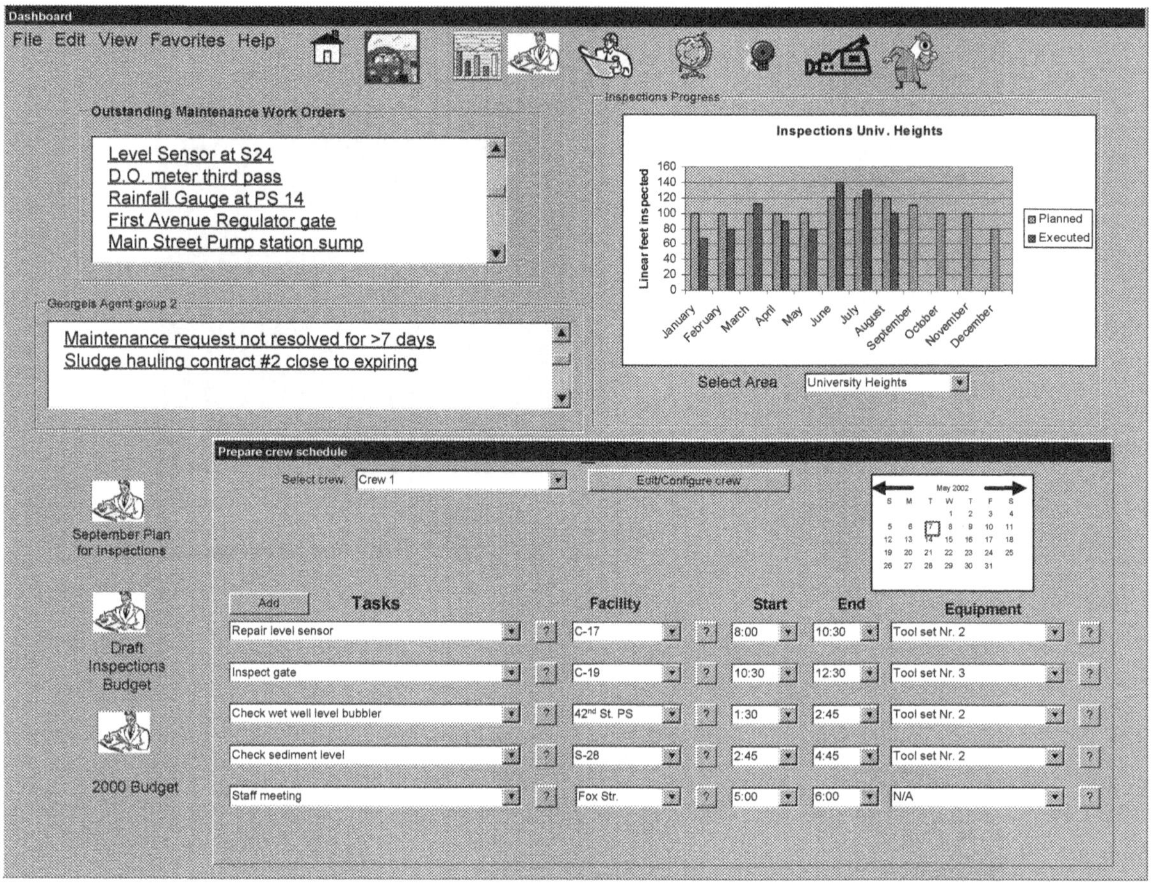

Figure 10-3. Example of Dashboard on Tactical Level.

Figure 10-3 includes the type of information that supervisor may want to keep track of, including the status of the inspections, and the schedule for the maintenance crews. The dashboard also includes links to documents that are important to the supervisor, such as budgets.

And, finally, Figure 10-4 shows what a dashboard might look like for a manager who is interested in more strategic information. Information on this dashboard comes from several different sources, including the project management system (tracking the status of budgets for different projects). Note that there is also information on a tactical level (monthly combined sewer overflow, or CSO, report) that might be on interest as well.

In the lower left corner, links are provided to documents that a manager may find helpful or interesting. These documents would be updated by different people, and would simply made available to the manager through a dashboard. The viewer used for such documents might be word processing software, such as Word for Windows or, even more appropriately, Adobe Acrobat Reader.

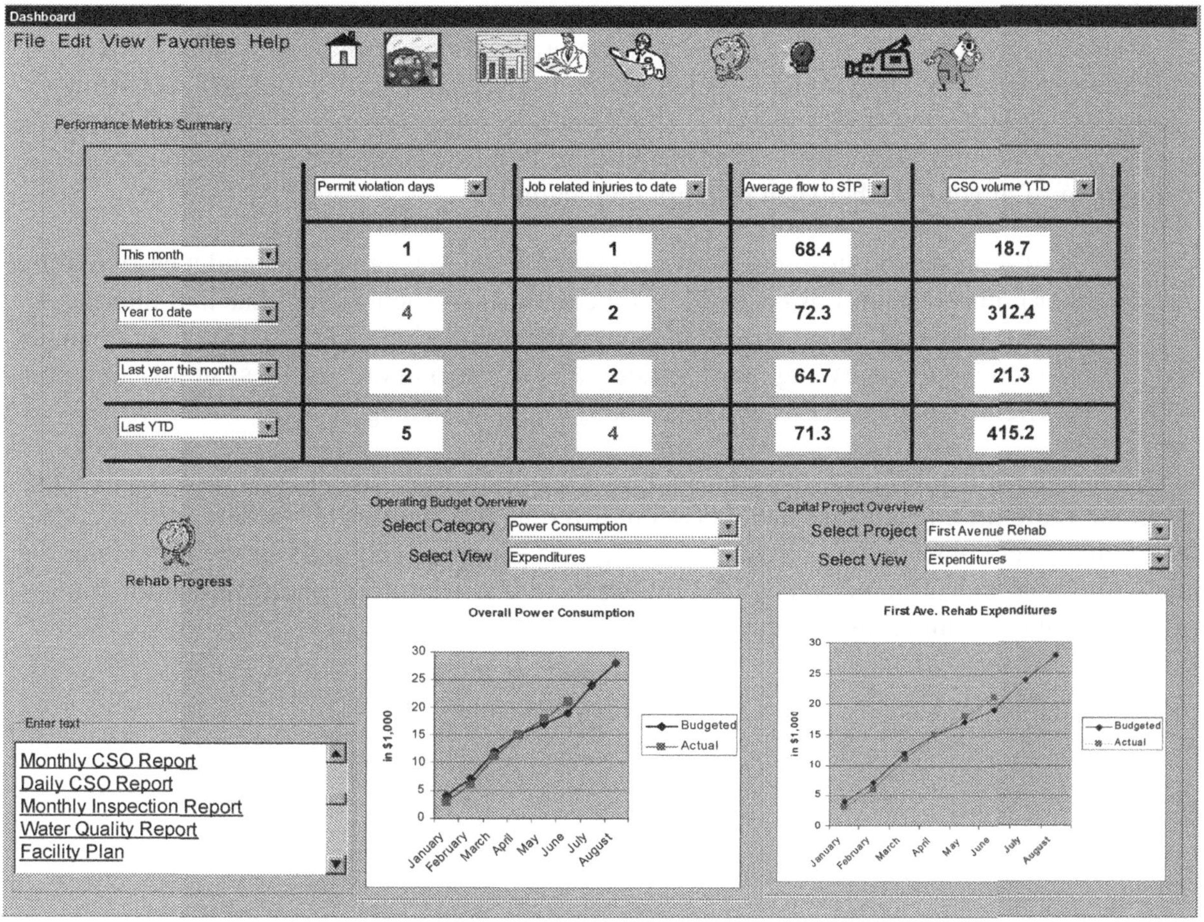

Figure 10-4. Example of Dashboard on Strategic Level.

The examples shown above are intended only to illustrate how information that is synthesized from different sources might be used to provide support on different levels of the organization, to people with different areas of interest. Although the content of information varies, the presentation uses several standard components such as graphs, map views, etc.

10.5 What Type of Information Could DSS Handle?

Because of different reasons including history, politics, and regional characteristics and priorities, there will be significant differences in the degree of automation achieved at different agencies. Some agencies in the United States have been much more aggressive than others in implementing automation and today have advanced and sophisticated systems. Other agencies, however, have done very little in terms of automation and are only now starting to implement systems. Therefore, the data available in the existing primary applications may be very different from agency to agency.

It is reasonable to assume that the agencies that are lagging behind today will eventually broaden the scope of automation. Therefore, it is important that the DSS should be able to accommodate the information that they may one day start to collect.

10.6 Secondary Benefits: Subscriptions

The integration of IT systems across the enterprise is a complex effort that demands significant resources that may require collaboration between many different groups within an organization. Reluctant to start a task of such magnitude, many organizations pursue integration in a piece meal fashion by point-to-point integration. Point-to-point integration includes development of direct links between primary applications.

A list of typical point-to-point examples is provided in Chapter 6.0, and a case might be made for other point-to-point links as well. As the number of such direct links and interfaces grows, however, the overall management and support of these links can become very difficult. Every time something changes (e.g., new version of software) in one of the systems, a number of links may be affected. This can create a nightmare for an IT manager.

The ì publish/subscribeî integration model using the centralized DSS data can also be used to facilitate the exchange of data between different applications. Therefore, instead of developing direct links between applications, the DSS database and tools serve as the main vehicle for communicating between different applications. Therefore, each application only needs to maintain a single link (to the DSS database).

CHAPTER 11.0

TECHNICAL REQUIREMENTS

In addition to satisfying the functional user requirements, a DSS should also meet some technical requirements. Although user requirements address *what* a DSS should do for the user, the technical requirements describe *how* a DSS should do it properly. The basic generic technical requirements for a DSS include the following:

1. Compatibility with the IT environment and applications within the organization; most organizations will have an IT Plan that defines the organizationís vision for IT. The DSS should fit well into that vision;
2. Open framework for adding modules: DSS will most likely be implemented in phases, and there has to be flexibility to add modules fairly easily;
3. A multi-user system that runs over the network: a DSS will need to draw information from many sources and possibly serve a number of different people in different departments;
4. System administration, management, and security; the DSS should not introduce unreasonable burden on the IT staff and resources. It should make life easier, not more complicated.

Because there will be differences in IT environments between different organizations, flexibility (ability to work with different types of applications) is a key technical property of a successful DSS.

Most large municipalities develop long-range strategic plans for IT. As part of the strategic IT planning process, it is necessary to define the main business objectives and lay out an IT vision for the organization. IT strategic plans define several aspects of the overall IT environment, and provide a basis for the procurement, development, and implementation of software applications and tools.

Therefore, because the strategic IT plan will define the specifics (including the individual applications), it is also the best starting point for planning the synthesis and integration: the development and implementation of a DSS. As users consider the requirements for a DSS, the strategic IT plan will also be a critical source of information about the overall technology direction that was selected for IT, the selected standards, and the software systems and applications that are planned for development and implementation.

If the organization has a strategic IT plan, this provides the guidance for defining the specific technical requirements for a DSS. It is even better if the considerations regarding software integration can be incorporated into the development of strategic plans, and thus the issues of integration are considered early.

If the organization does not have a strategic IT plan, it will be necessary to conduct some analyses to determine the technical requirements for a DSS. These analyses include a definition of the business objectives, a survey of existing (legacy) applications, and a plan for integration.

CHAPTER 12.0

CONCLUSIONS AND RECOMMENDATIONS

12.1 Conclusions

This research project was structured into three parts, resulting in three reports that covered the following issues:

- What is the state-of-the-art of DSS in wastewater industry?
- What is private industry doing about DSS?
- How can the decision-making process be described and improved upon?
- How can this approach and technology be successfully applied to wastewater agencies?
- What are the components of a DSS, and what type of functionality could such a system provide to wastewater professionals?

The main conclusions of this research, summarized from the three project reports, are as follows:

- To improve the outcome of the decision-making process (decisions), it is necessary to understand and improve the decision-making process. Decision analysis provides the conceptual framework and specific tools for describing the decision-making process and gives us methods and a standard notation for presenting decision-making processes in a structured way.
- Defining the objectives is a critical step in the development and analysis of decisions and DSS.
- The private (ìparallelî) industry perceives system integration and decision support to be key issues, and has devoted very significant resources into modeling and understanding their business processes.
- Although there are some similarities between the private industry and wastewater utilities, the differences between them remain significant. Although some valuable lessons can be learned from research into the practices of private industries, users must apply these lessons with caution and consider that the business environment and conditions in wastewater utilities is different.
- Decision-making process relies on information, and currently information about different aspects of wastewater utilitiesí business is scattered in a number of different systems. To improve the decision-making process, it is necessary to provide more effective methods and tools for accessing the information throughout the enterprise (organization). Therefore, issues of system integration are very closely linked to decision support.

- The IT industry does not today offer a complete and generic ì packagedî solution that fully addresses the problems of decision support.
- Wastewater agencies/utilities include a wide variety of business processes that range from process control to strategic issues. While private industry has devoted significant resources to develop reference business process models (SCOR, TOM), such a reference model is not available for wastewater industry.
- Modeling of business processes is important, as it defines how the organization goes about accomplishing its mission. This report discusses several methods and off-the-shelf tools for the development of business process models.
- As part of this project, an initial set of business process descriptions for a wastewater agency was developed and organized using a tool based on use case approach.
- Wastewater professionals and managers of wastewater utilities must make decisions that often require consideration of complex scientific analyses (related to environmental impact) combined with the need to take into account a number of different stakeholders and operate in a politically complex environment. Although decision support tools can often provide significant benefits in such situations, it is important to recognize the complexity of the problem.

As a final note, it is important to emphasize that the situation related to DSS is dynamic and that new projects, tools, ideas, standards, and methods related to decision support and system integration will continue to appear. The shelf life of a report that deals with issues that relate to IT is precarious and short. The world of IT is a rapidly moving target. This report attempts to provide a quick glimpse into a broad and complex set of issues and problems. The result of this research does not by any means provide the final word on decision support, but instead will hopefully provide a useful introduction and guide to wastewater professionals that are keen to explore these exciting and challenging issues.

12.2 Recommendations

The researchers recommend that individual agencies consider the benefits of system integration and DSS on their performance. WERF and its utility members should continue monitoring the evolution of DSS and enterprise computing (integration) efforts and systems. Also, because of the fast pace of changes in IT, WERF should consider an update to this study in 2 to 3 years. Finally, the researchers note that a collaborative pilot project, conducted by several utilities, could be an optimal mechanism for developing the basic components of a DSS framework that would focus on the needs of the wastewater industry.

References

Armour F., Miller, G. *Advanced Use Case Modeling*. New York: Addison Wesley, 2001.

American Water Works Association Research Foundation. AWWARF committee report: UCA (Utility Communications Architecture) for the water industry, *J. Am. Water Works Assoc.* June 2001, 68ñ81.

Baron, J. *Thinking and Deciding.* New York: Cambridge University Press, 2000.

Baron L., Aichiche, S., Balazinski, M. Fuzzy decision support system knowledge base generation using a genetic algorithm, *Int. J. Approx. Reason.*2001, *28*, 125ñ148.

Brueck, T.M. *Developing and Implementing a Performance Measurement System: Vol. 1*, WERF Final Report 99-WWF-7. Alexandria, VA: Water Environment Research Foundation, 2000.

Clemen, R. *Making Hard Decisions - An Introduction to Decision Analysis*. Pacific Grove, CA: Duxbury Press/Thomson, 1996.

Cockburn, A., *Writing Effective Use Cases*. Upper Saddle River, NJ: Addison Wesley/Pearson, 2001.

Comas J., Dzeroski, S., Gilbert, K., Rodriguez-Roda, I., Sanchez-Marre, M. Knowledge discovery by means of inductive methods in wastewater treatment plant data, *AI Commun.* 2001, 14, 45ñ62.

Cortes U., Sanchez-Marre, M., Sanguesa, R., Comas, J., Rodriguez-Roda, I., Poch, M., Riano, D. Knowledge management in environmental decision support systems, *AI Commun*, 2001, *14*, 3ñ 12.

Eisenhardt, P., Waltrip, G. D. *Improving Wastewater Treatment Plant Operations Efficiency and Effectiveness,* WERF Final Report 97-CTS-1. Alexandria, VA: Water Environment Research Foundation, 1999.

Humphrey, W. *Managing the Software Process*. New York: Addison-Wesley, 1990.

Jacobson I., Formalizing use-case modeling, *JOOP, J. Object-Oriented Program.* June 1995, 10ñ 14.

Keeney, R. *Value Focused Thinking.* Boston: Harvard University Press, 1992.

King County Wastewater Treatment Division. *Computer Systems Planning Study Final Report*. King County, Wash., Seattle, 2002. Available at http://dnr.metrokc.gov/wtd/compstudy/. Accessed 8/4/2004.

Love, N. G., Bott, C.B. *A Review and Needs Survey of Upset Early Warning Devices*, WERF Final Report 99-WWF-2. Alexandria, VA: Water Environment Research Foundation, 2000.

March, J. *A Primer on Decision Making; How Decisions Happen*. New York: Free Press/Simon & Schuster, 1994.

Oxenford, J. *Using the Utility Business Architecture to Become Competitive*, Report 90765. Denver, CO: American Water Works Association Research Foundation, 1998.

Parks, S. *The Utility Business Architecture: Designing for Change,* Report 90726. Denver, CO: AWWA Research Foundation, 1997.

Patrick, R., Rompala, J., Symkowski, A., Kingdom, W., Serpente, R., Freeman, N., Stevens, B., Koch, C., Kochaba, T. *Benchmarking Wastewater Operationsó Collection, Treatment, and Biosolids Management,* WERF Final Report 96-CTS-5. Alexandria, VA: Water Environment Research Foundation, 1997.

Rodriguez-Roda, I., Poch, M., Banares-Alcantara, R. Conceptual design of wastewater treatment plants using a design support system, *J. Chem. Technol. Biotechnol.* 2000, *75*, 73ñ81.

Rodriguez-Roda, I., Comas, J., Poch, M. Automatic knowledge acquisition from complex processes for the development of knowledge-based systems, *Ind. Eng. Chem. Res*. 2001, *40* (*15*), 3353ñ3360.

Sachs, O. *The Man Who Mistook his Wife for a Hat*. New York: Perennial Library, Harper & Row, 1987.

Scannell, E., Jones, M. SAP advances on Websphere turf, *InfoWorld*. June 20, 2003.

Skinner, D. *Introduction to Decision Analysis*. Gainesville, FL: Probabilistic Publishing, 1995.

Supply Chain Council, Inc. *Supply Chain Operations Reference Modeló SCOR Version 4.0ó an industry reference model*. Pittsburgh, PA: Supply Chain Council, Inc., August 2000.

Telecommunications operations mapó A telecom industry white paper. March 2000.

Wastewater Testing of North America. *On-line Monitoring to Control Transients in Wastewater Treatment.* WERF Final Report 92-OPW-1. Alexandria, VA: Water Environment Research Foundation, 1994.

WASTEWATER UTILITY

Alabama
Montgomery Water Works & Sanitary Sewer Board

Alaska
Anchorage Water & Wastewater Utility

Arizona
Gila Resources

Glendale, City of, Utilities Department

Mesa, City of

Peoria, City of

Pima County Wastewater Management

Arkansas
Little Rock Wastewater Utility

California
Calaveras County Water District

Central Contra Costa Sanitary District

Contra Costa Water District

Crestline Sanitation District

Delta Diablo Sanitation District

Dublin San Ramon Services District

East Bay Dischargers Authority

East Bay Municipal Utility District

El Dorado Irrigation District

Fairfield-Suisun Sewer District

Fresno Department of Public Utilities

Irvine Ranch Water District

Las Virgenes Municipal Water District

Livermore, City of

Lodi, City of

Los Angeles, City of

Los Angeles County, Sanitation Districts of

Napa Sanitation District

Orange County Sanitation District

Palo Alto, City of

Riverside, City of

Sacramento Regional County Sanitation District

San Diego Metropolitan Wastewater Department, City of

San Francisco, City & County of

San Jose, City of

Santa Barbara, City of

Santa Rosa, City of

South Bayside System Authority

South Coast Water District

South Orange County Wastewater Authority

Sunnyvale, City of

Union Sanitary District

West Valley Sanitation District

Colorado
Boulder, City of

Colorado Springs Utilities

Littleton/Englewood Water Pollution Control Plant

Metro Wastewater Reclamation District, Denver

Connecticut
The Mattabasset District

New Haven, City of, WPCA

District of Columbia
District of Columbia Water & Sewer Authority

Florida
Broward, County of

Fort Lauderdale, City of

Gainesville Regional Utilities

JEA

Miami-Dade Water & Sewer Authority

Orange County Utilities Department

Orlando, City of

Reedy Creek Improvement District

Seminole County Environmental Services

St. Petersburg, City of

Stuart Public Utilities

Tallahassee, City of

Tampa, City of

Toho Water Authority

West Palm Beach, City of

Georgia
Atlanta Department of Watershed Management

Augusta, City of

Clayton County Water Authority

Cobb County Water System

Columbus Water Works

Fulton County

Gwinnett County Department of Public Utilities

Savannah, City of

Hawaii
Honolulu, City and County of

Illinois
American Bottoms Wastewater Treatment Plant

Greater Peoria Sanitary District

Kankakee River Metropolitan Agency

Metropolitan Water Reclamation District of Greater Chicago

North Shore Sanitary District

Wheaton Sanitary District

Iowa
Ames, City of

Cedar Rapids Wastewater Facility

Des Moines Metro Wastewater Reclamation Authority

Iowa City, City of

Kansas
Johnson County Unified Wastewater Districts

Unified Government of Wyandotte County/Kansas City, City of

Kentucky
Louisville & Jefferson County Metropolitan Sewer District

Louisiana
Sewerage & Water Board of New Orleans

Maine
Bangor, City of

Portland Water District

Maryland
Anne Arundel County Bureau of Utility Operations

Howard County Department of Public Works

Washington Suburban Sanitary Commission

Massachusetts
Boston Water & Sewer Commission

Upper Blackstone Water Pollution Abatement District

Michigan
Ann Arbor, City of

Detroit, City of

Holland Board of Public Works

Lansing, City of

Owosso Mid-Shiawassee County WWTP

Saginaw, City of

Wayne County Department of Environment

Wyoming, City of

Minnesota
Rochester, City of

Western Lake Superior Sanitary District

Missouri
Independence, City of

Kansas City Missouri Water Services Department

Little Blue Valley Sewer District

Nebraska
Lincoln Wastewater System

Nevada
Henderson, City of

New Jersey
Bergen County Utilities Authority

Passaic Valley Sewerage Commissioners

New York
New York City Department of Environmental Protection

Rockland County Solid Waste Management Authority/ Sewer District

North Carolina
Charlotte/Mecklenburg Utilities

Durham, City of

Metropolitan Sewerage District of Buncombe County

Orange Water & Sewer Authority

Ohio
Akron, City of

Butler County Department of Environmental Services

Columbus, City of

Metropolitan Sewer District of Greater Cincinnati

Northeast Ohio Regional Sewer District

Summit, County of

Oklahoma
Tulsa, City of

Oregon
Clean Water Services

Eugene/Springfield Water Pollution Control

Water Environment Services

Pennsylvania
Philadelphia, City of

University Area Joint Authority, State College

South Carolina
Charleston Commissioners of Public Works

Mount Pleasant Waterworks & Sewer Commission

Spartanburg Sanitary Sewer District

Tennessee
Cleveland, City of

Knoxville Utilities Board

Murfreesboro Water & Sewer Department

Nashville Metro Water Services

Texas
Austin, City of

Dallas Water Utilities

Denton, City of

El Paso Water Utilities

Fort Worth, City of

Gulf Coast Waste Disposal Authority

Houston, City of

San Antonio Water System

Trinity River Authority

Utah
Salt Lake City Corporation

Virginia
Alexandria Sanitation Authority
Arlington, County of
Fairfax County Virginia
Hampton Roads Sanitation District
Henrico, County of
Hopewell Regional Wastewater Treatment Facility
Loudoun County Sanitation Authority
Lynchburg Regional WWTP
Prince William County Service Authority
Richmond, City of
Rivanna Water & Sewer Authority

Washington
Edmonds, City of
Everett, City of
King County Department of Natural Resources
Seattle Public Utilities
Sunnyside, Port of
Yakima, City of

Wisconsin
Green Bay Metro Sewerage District
Madison Metropolitan Sewerage District
Milwaukee Metropolitan Sewerage District
Racine, City of
Sheboygan Regional Wastewater Treatment
Wausau Water Works

Australia
South Australian Water Corp.
Sydney Water Corp.
Water Corp. of Western Australia

Canada
Greater Vancouver Regional District
Toronto, City of, Ontario
Winnipeg, City of, Manitoba

Mexico
Servicios de Agua y Drenaje de Monterrey, I.P.D.

New Zealand
Watercare Services Limited

Singapore
Singapore Public Utilities Board

United Kingdom
Yorkshire Water Services Limited

STORMWATER UTILITY

California
Los Angeles, City of, Department of Public Works
Monterey, City of
San Francisco, City & County of
Santa Rosa, City of
Sunnyvale, City of

Colorado
Boulder, City of

Georgia
Griffin, City of

Iowa
Cedar Rapids Wastewater Facility
Des Moines Metro Wastewater Reclamation Authority

Kansas
Overland Park, City of

Kentucky
Louisville & Jefferson County Metropolitan Sewer District

Maine
Portland Water District

Minnesota
Western Lake Superior Sanitary District

North Carolina
Charlotte, City of, Stormwater Services

Pennsylvania
Philadelphia, City of

Tennessee
Chattanooga Storm Water Management

Washington
Seattle Public Utilities

STATE

Arkansas Department of Environmental Quality
Fresno Metropolitan Flood Control District, Calif.
Urban Drainage & Flood Control District, Colo.

CORPORATE

ADS Environmental Services
The ADVENT Group Inc.
Alan Plummer & Associates
Alden Research Laboratory
Aqua-Aerobic Systems, Inc.
Aquateam–Norwegian Water Technology Centre A/S
BaySaver Inc.
BioVir Laboratories, Inc.
Black & Veatch
Boyle Engineering Corporation
Brown & Caldwell
Burns & McDonnell
CABE Associates Inc.
The Cadmus Group
Camp Dresser & McKee Inc.
Carollo Engineers Inc.
Carpenter Environmental Associates Inc.
CDS Technologies Inc.
Chemtrac Systems Inc.
CH2M HILL
Damon S. Williams Associates, LLC
David L. Sheridan, P.C.
Dewling Associates, Inc.
Earth Tech Inc.
Eco-Matrix
Ecovation
Environmental Engineers International
EMA Inc.
The Eshelman Company Inc.
Finkbeiner, Pettis, & Strout Inc.
Freese & Nichols, Inc.
ftn Associates Inc.
Fuss & O'Neill, Inc.
Gannett Fleming Inc.
Golder Associates Inc.
Greeley and Hansen LLC
Hazen & Sawyer, P.C.
HDR Engineering Inc.
HNTB Corporation
HydroQual Inc.
Infilco Degremont Inc.
Ingersoll-Rand Energy Systems
Insituform Technologies Inc.
Institute for Environmental Technology & Industry, Korea
Jacobson Helgoth Consultants Inc.
Jason Consultants Inc.
Jordan, Jones, & Goulding Inc.
KCI Technologies Inc.
Kelly & Weaver, P.C.
Kennedy/Jenks Consultants
Komline Sanderson Engineering Corporation
Lawler, Matusky & Skelly Engineers, LLP
Limno-Tech Inc.
Lombardo Associates Inc.
Malcolm Pirnie Inc.
McKim & Creed
MEC Analytical Systems Inc.
Metcalf & Eddy Inc.
MWH
New England Organics
Odor & Corrosion Technology Consultants Inc. (OCTC)
Oswald Green, LLC
PA Government Services Inc.
Parametrix Inc.
Parsons
Post, Buckley, Schuh & Jernigan
R&D Engineering/Conestoga Rover & Associates
The RETEC Group
RMC, Inc. (Raines, Melton & Carella)
R.M. Towill Corporation
Ross & Associates Ltd.
Royce Technologies
SAIC Maritime Technical Group
Short Elliott Hendrickson, Inc.
Stantec Consulting Group, Inc.
Stormwater Management, Inc.
Synagro Technologies, Inc.
Tetra Tech Inc.
Trojan Technologies Inc.
URS Corporation
Veolia Water NATC
Wade-Trim Inc.
Weston Solutions Inc.
Woodard & Curran
WRc/D&B, LLC
WWETCO, LLC
Zoeller Pump Company

INDUSTRY

American Electric Power
ChevronTexaco Energy Research & Technology Company
The Coca-Cola Company
Dow Chemical Company
DuPont Company
Eastman Kodak Company
Eli Lilly & Company
Merck & Company Inc.
ONDEO Services
Procter & Gamble Company
PSEG Services Corp.
RWE Thames Water Plc
Severn Trent Services Inc.
United Water Services LLC

Note: List as of 7/1/04

Board of Directors

Research Council

Stormwater Technical Advisory Committee

WERF Product Order Form

As a benefit of joining the Water Environment Research Foundation, subscribers are entitled to receive one complimentary copy of all final reports and other products. Additional copies are available at cost (usually $10). To order your complimentary copy of a report, please write "free" in the unit price column. WERF keeps track of all orders. If the charge differs from what is shown here, we will call to confirm the total before processing.

Name ______________________ Title ______________________

Organization ______________________

Address ______________________

City ______________ State ______ Zip Code ______ Country ______

Phone ______________ Fax ______________ Email ______________

Stock #	Product	Quantity	Unit Price	Total
			Postage & Handling	
			VA Residents Add 4.5% Sales Tax	
			Canadian Residents Add 7% GST	
			TOTAL	

Method of Payment:

(All orders must be prepaid.)

- ❑ Check or Money Order Enclosed
- ❑ Visa ❑ Mastercard ❑ American Express

Account No. ______________ Exp. Date ______

Signature ______________________

Shipping & Handling:

Amount of Order	United States	Canada & Mexico	All Others
Up to but not more than:	Add:	Add:	Add:
$20.00	$5.00*	$8.00	50% of amount
30.00	5.50	8.00	40% of amount
40.00	6.00	8.00	
50.00	6.50	14.00	
60.00	7.00	14.00	
80.00	8.00	14.00	
100.00	10.00	21.00	
150.00	12.50	28.00	
200.00	15.00	35.00	
More than $200.00	Add 20% of order	Add 20% of order	

*minimum amount for all orders

Note: Please make checks payable to the Water Environment Research Foundation.

To Order (Subscribers Only):

Log on to www.werf.org and click on the "Product Catalog."

Phone: (703) 684-2470
Fax: (703) 299-0742.

WERF
Attn: Subscriber Services
635 Slaters Lane
Alexandria, VA 22314-1177

To Order (Non-Subscribers):

Non-subscribers may be able to order WERF publications either through WEF (www.wef.org) or IWAP (www.iwapublishing.com). Visit WERF's website at www.werf.org for details.

IWA Publishing's authorised EU representative for General Product Safety Regulations is Diane D'Arras, 15 rue Duret, 75116 Paris, France, e-mail: safety@iwap.co.uk.

Printed and bound by CPI Group (UK) Ltd, Croydon, CR0 4YY
12/05/2026
02108912-0001